Advances in
MICROBIAL ECOLOGY
Volume 3

ADVANCES IN MICROBIAL ECOLOGY

Sponsored by International Commission on Microbial Ecology,
a unit of International Association of Microbiological Societies
and the Division of Environmental Biology of the
International Union of Biological Societies

A Continuation Order Plan is available for this series. A continuation order will bring delivery of each new volume immediately upon publication. Volumes are billed only upon actual shipment. For further information please contact the publisher.

Advances in
MICROBIAL ECOLOGY

Volume 3

Edited by

M. Alexander

Cornell University
Ithaca, New York

PLENUM PRESS · NEW YORK AND LONDON

The Library of Congress cataloged the first volume of this title as follows:

Advances in microbial ecology. v. 1—
  New York, Plenum Press c1977—
    v. ill. 24 cm.
  Key title: Advances in microbial ecology, ISSN 0147-4863

  1. Microbial ecology—Collected works.
QR100.A36   576′.15   77-649698

Library of Congress Catalog Card Number 77-649698

ISBN-13: 978-1-4615-8281-6   e-ISBN-13: 978-1-4615-8279-3
DOI: 10.1007/978-1-4615-8279-3

Contributors

G. H. W. Bowden, MRC Dental Epidemiology Unit, London Hospital Dental School, London, England. Present address: Department of Oral Biology, Faculty of Dentistry, University of Manitoba, Winnipeg, Canada

D. L. Collins-Thompson, Bureau of Microbial Hazards, Health Protection Branch, Ottawa, Canada K1A 0L2

D. C. Ellwood, Microbiological Research Establishment, Porton Down, England

I. R. Hamilton, Department of Oral Biology, Faculty of Dentistry, University of Manitoba, Winnipeg, Canada

A. Hurst, Bureau of Microbial Hazards, Health Protection Branch, Ottawa, Canada K1A 0L2

E. Padan, Department of Microbiological Chemistry, The Hebrew University-Hadassah Medical School, P.O.B. 1172, Jerusalem, Israel

Meyer J. Wolin, Division of Laboratories and Research, New York State Department of Health, Albany, New York 12201, U.S.A.

Preface

We are most gratified by the response to the initiation of this series of volumes presenting recent developments and new concepts in microbial ecology. Favorable reactions have been expressed in both oral and written communication, and *Advances in Microbial Ecology* thus seems to be providing a worthwhile outlet in a rapidly growing field of microbiology and environmental sciences.

The growing importance of microbial ecology is evident in many ways. University personnel are expanding their programs and increasing the number of research topics and publications. Substantial numbers of industrial scientists have likewise entered this field as they consider the microbial transformation of chemicals in waters and soils and the effects of synthetic compounds on natural microbial communities. Agricultural, medical, dental, and veterinary practitioners and scientists have also been increasing their activity in microbial ecology because of the importance of the discipline to their own professions. In addition, governmental agencies have expanded regulatory and research activities concerned with microbial ecology owing to the importance of information and regulations focused on the interactions between microorganisms in nature and particular environmental stresses.

The present volume maintains the approach formulated originally by the International Commission on Microbial Ecology. The reviews thus deal with both basic and applied microbiology and are concerned with aquatic, oral, rumen, and food ecosystems. Moreover, diverse groups of organisms are the subject of the several reviews, and the approaches of the authors differ substantially according to the professional interests, requirements, and scope appropriate for the various disciplines. An international group of authors likewise contributes to the present volume. It is the hope of the Editorial Board that future volumes will continue to reflect this breadth: both basic and applied topics, diverse ecosystems, various groups of microorganisms, and an international group of authors.

In this light, we encourage our colleagues in various aspects of microbial ecology and environmental microbiology to submit titles and outlines for prospective reviews to members of the Editorial Board. Now that the series is well established, we welcome unsolicited manuscripts but hope that prospective authors will consult us before preparing full manuscripts so that an assessment of the approach and relevancy of the manuscript can be made. We thus hope that the *Advances* will continue to serve a useful function and will be a vehicle not only for the dissemination of current information but also for the promotion of the further development of microbial ecology.

The Editor and Editorial Board are appointed by the International Commission on Microbial Ecology for fixed terms. Moshe Shilo has now completed his term on the Editorial Board, and the Commission, the Editor, and his colleagues on the Editorial Board express to him their sincere gratitude for his cooperation, professional advice, and willingness to help in initiating *Advances in Microbial Ecology*.

M. Alexander, Editor
T. Rosswall
M. Shilo
H. Veldkamp

Contents

Chapter 1
Impact of Facultatively Anaerobic Photoautotrophic Metabolism
on Ecology of Cyanobacteria (Blue-Green Algae)
E. Padan

Chapter 2
The Rumen Fermentation: A Model for Microbial Interactions in Anaerobic Ecosystems
Meyer J. Wolin

Chapter 3
Food as a Bacterial Habitat
A. Hurst and D. L. Collins-Thompson

Chapter 4

Microbial Ecology of the Oral Cavity

G. H. W. Bowden, D. C. Ellwood, and I. R. Hamilton

Impact of Facultatively Anaerobic Photoautotrophic Metabolism on Ecology of Cyanobacteria (Blue-Green Algae)

E. PADAN

1. Introduction

This review attempts to evaluate the ecological importance of a recently discovered physiological character in cyanobacteria (blue-green algae), anaerobic photoautotrophic metabolism, by considering the possible expression of the character in the natural environment and its selective value to the organism. In striving for clarity and emphasis, this analysis must concentrate on major factors and may be oversimplified. Nevertheless, as in literature, theater, and other fields of human endeavor, if the ideas put forward are understood and elicit a response, the effort will have been worthwhile.

Cohen *et al.* (1975a) demonstrated that a cyanobacterium, *Oscillatoria limnetica*, is capable of anoxygenic CO_2 photoassimilation with sulfide as an electron donor in a photosystem-I-driven reaction. This reaction has since been thoroughly investigated (Cohen *et al.*, 1975b; Oren *et al.*, 1977; Oren and Padan, 1978; Belkin and Padan, 1978). More recently, additional cyanobacteria were shown to possess this physiological character (Castenholtz, 1976, 1977; Garlick *et al.*, 1977).

E. PADAN • Department of Microbiological Chemistry, The Hebrew University–Hadassah Medical School, P.O.B. 1172, Jerusalem, Israel.

2. Aquatic Systems with Alternating Photoaerobic–Photoanaerobic Conditions

A light-dependent sulfide-utilizing metabolic mechanism is most likely to be expressed in sulfide-rich ecosystems with light penetration. As sulfide is readily oxidized by O_2, the maintenance of a particular sulfide concentration in a habitat is determined by the ambient O_2 tension. Thus, a gradient of O_2 is often accompanied by a gradient of sulfide under natural conditions. Furthermore, sulfide accumulation occurs frequently under anaerobic conditions when biological and chemical activities decompose organic material and/or reduce oxidized sulfur compounds. With the enrichment in sulfide, there is a sharp discontinuity in the redox potential between the aerobic and sulfide-containing habitats. The redox potential of the latter can reach -200 mV and even lower values (Baas Becking and Wood, 1955; Fenchel and Riedl, 1970; Fenchel, 1971; Gest, 1972; Pfennig, 1975; Cohen et al., 1977a). The photic sulfide-rich semianaerobic to anaerobic situation is very characteristic of transparent water bodies which, with limited mixing, can become closed systems with respect to air.

2.1. Hot Sulfur Springs

Hot sulfur springs provide a familiar example of a gradient in sulfide concentration. Castenholtz (1976) described alkaline and neutral hot springs in New Zealand, with particular attention to some from the central volcanic zone of North Island with up to 2 mM sulfide at the source and to other springs from southwest Iceland with lower sulfide concentrations of up to 0.3 mM. In the Yellowstone area of the United States (Castenholtz, 1977) (Fig. 1), the source waters of the Upper Terrace of Mammoth Springs have pH values ranging between 6.2 and 6.8 with a mean sulfide concentration of 56 μM and a maximum concentration of 0.13 mM. Acidic hot springs containing sulfide are widely distributed (Castenholtz, 1969). As sulfide is readily oxidized by O_2, the sulfide concentration in hot springs decreases with distance from the source; there is a decrease in temperature with a concomitant increase in O_2 content and pH (Castenholtz, 1977; Fig. 1). Hence, sulfur springs provide a gradient of conditions from the photoanaerobic sulfide-rich source waters to the photoaerobic waters downstream.

2.2. Stratified Lakes

The photoanaerobic sulfide-rich situation is exemplified in many lakes which undergo thermal stratification. [The limnological terminology is that of Hutchinson (1967).] The stratification pattern in lakes varies with different latitudes, altitudes, lake depths, and solutes. However, the summer stratification

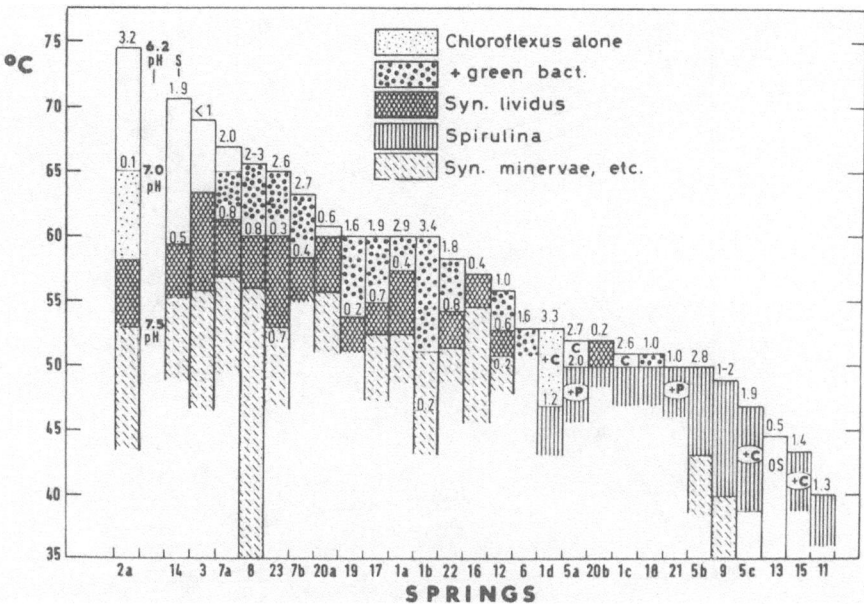

Figure 1. Predominant phototrophs in various temperature zones of springs in the Upper Terrace, Mammoth (Yellowstone National Park, U.S.A.). Each vertical rectangle represents a spring with the source at the top, declining in temperature (toward base) as in the drainway, and along the edges and vertically in hot pools. The sulfide concentration in mg/liter is indicated for each source and also at some species borders at lower temperatures. The pH range at the sources was from 6.2 to 6.7 except for nos. 14 and 11 (pH 6.8), no. 20b (6.9), and no. 20a (7.0). The specific conductance (μmhos/cm, 25°C) of these waters ranged from about 2200 to 2600. The species key: C, *Chromatium* sp.; P, *Phormidium* sp.; Os, *Oscillatoria* sp. The blank areas near some sources lack phototrophs but may include nonphotosynthetic bacteria. From Castenholtz (1977).

cycle in temperate lakes may serve as an example of the stratification process. After the spring overturn, warming of the upper water layers yields an upper layer (epilimnion) of less dense, freely circulating waters, an intermediate layer (metalimnion) with a temperature gradient (thermocline), and a noncirculating colder, denser bottom layer (hypolimnion). The thermal gradient causes a density stratification resulting in isolation of the hypolimnion. In shallow eutrophic lakes, photoanaerobic conditions may develop both in the stagnant water layer and in the mud. Density stratification in lakes may also be due to a chemical gradient (chemocline) (Hutchinson, 1967; Cohen *et al.*, 1977a,b) (Fig. 2).

Inherent to the physical parameters governing the lake system, the stratified situation is very often unstable with time. Aerobic and anaerobic conditions alternate in accordance with changes in the stratification. A few lakes do not

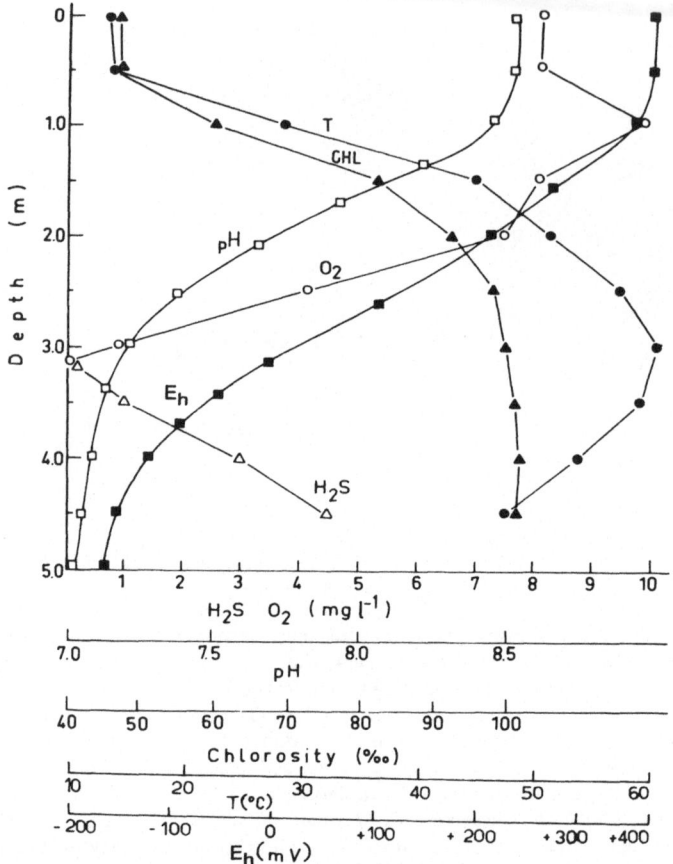

Figure 2. Vertical distribution of temperature, chlorosity, H_2S, O_2, pH, and E_h at height of stratification, April 4, 1974, in the Solar Lake (Israel). From Cohen *et al.* (1977a).

undergo overturn (amictic lakes); however, most lakes mix partly (meromictic lakes) or completely (holomictic lakes) once or twice yearly. In lower latitudes, mixing may occur more frequently (oligomictic lakes). In the equatorial region's shallow eutrophic lakes undergoing daily overturn (polymictic lakes), there is a daily cycle of stratification and holomixis with daily fluctuations in O_2 tensions (Baxter *et al.*, 1965; Talling *et al.*, 1973; Viner and Smith, 1973; Ganf and Viner, 1973; Ganf and Horne, 1975; Reynolds and Walsby, 1975; Greenwood, 1976).

The O_2 regime in Lake George (Uganda), a polymictic lake, has been thoroughly studied (Burgis *et al.*, 1973; Viner and Smith, 1973; Ganf and Viner, 1973; Ganf, 1974a,b; Ganf and Horne, 1975; Greenwood, 1976) (Fig. 3). At dawn the water column of Lake George is isothermal; from 10.00 hr onward there is a progressive build-up of thermal stratification which breaks down in the evening, so that at 18.00 hr the water column is isothermal again (Fig. 3a). There is a stratification in O_2 tension corresponding to the thermostratification of the water column (Fig. 3b). With the frequent fluctuations in O_2 tension imposed by this daily cycle, O_2 levels fall below saturation value, but usually not to the anaerobic level. However, Ganf and Viner (1973) have shown that the O_2 budget of the water column is very delicately balanced. The diurnal changes in conditions may cause anaerobiosis and resultant fish mortality. The mean O_2 content of the water column during the night is 13 g/m^2. The mean O_2 consumption of the upper 5 cm of a 1-m^2 area of bottom mud is 5 g during the first hour of restored contact with O_2. If wave action at the sediment–water interface is sufficient to suspend this surface mud in the water column even for an hour, a

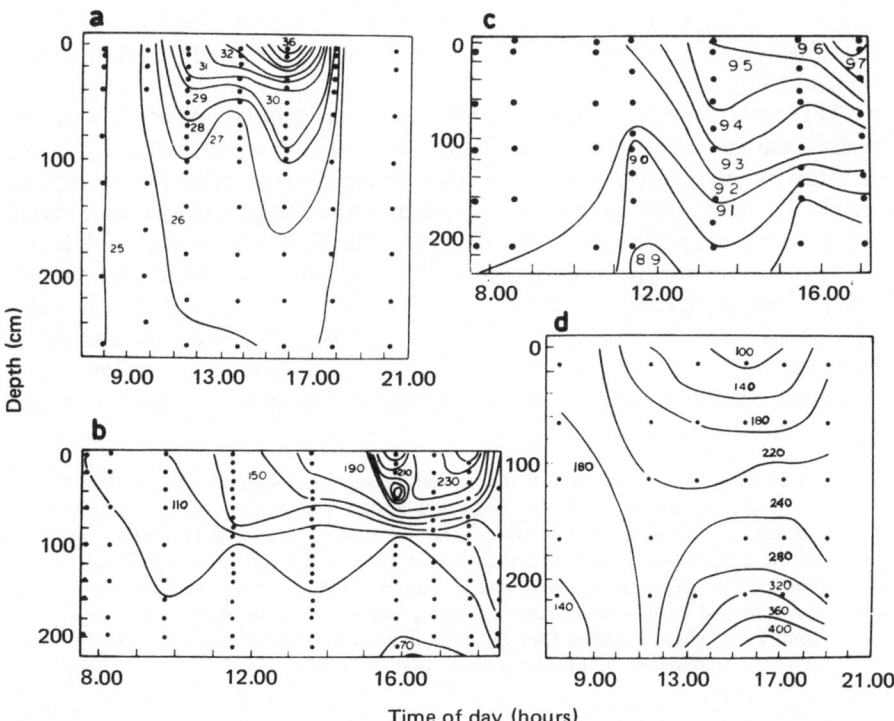

Figure 3. Diurnal vertical distribution of temperature, oxygen, pH and chlorophyll a in Lake George (Uganda), March 26, 1968. a: Temperature–isopleth intervals are 1°C; dots mark the points at which measurements were taken. b: Percentage oxygen saturation–isopleth intervals are 1–10%. c: pH–isopleth intervals are 0,1. d: Chlorophyll a–isopleth intervals are 40 mg/m^3. From Ganf and Horne (1975).

significant proportion of the available O_2 in the water will be consumed. Sediment disturbance to 5 cm and greater depths is not uncommon, and if this sediment remains in suspension during the subsequent night, anoxia develops. Thus, whenever the thermocline persists for several days, the hypolimnion is completely deoxygenated, resulting in photoanaerobic conditions the following dawn.

2.3. Aquatic Systems with Photoaerobic to Dark Anaerobic Alternations

In many aquatic systems, as in polymictic lakes, anaerobic sulfide-rich conditions are often established during the night. These are separated in time from the daily aerobic photic conditions. However, an interval of photic anaerobiosis may be expected at dawn. Thus, the photosynthetic community is exposed to frequent anaerobic conditions. In stratified eutrophic lakes, fluctuating anaerobic conditions are not uncommon even in the epilimnion (Hutchinson, 1967; Fogg and Walsby, 1971; Sirenko, 1972; Reynolds and Walsby, 1975; Whitton and Sinclair, 1975).

The cycle described above is typical of very shallow bodies of water of several centimeters' depth, covering large areas of the world, i.e., sea marshes of estuaries, mangroves, and rice paddies. These systems, often rich in organic matter, undergo very marked O_2 tension fluctuations within a daily cycle (Singh, 1961; Fogg et al., 1973; Brock, 1973a; Reynolds and Walsby, 1975).

The shallow marshes behind the beach line of the Texas coast, with limited connections to the Gulf of Mexico, are an example (Odum, 1967). In these millions of acres of shallow polluted marine waters, the living community is compressed into a film which may be less than 10 cm deep. The shallowness of the water column amplifies the diurnal ranges of properties responding to the daily insolation cycle.

> Thus, the shallower the water, the greater becomes the diurnal range of temperature, oxygen, and pH. . . . There is a hyperbolic relation of these ranges decreasing with depth. High ranges produce almost anaerobic conditions at night, oxygen being used as fast as it diffuses into the film. Since films cannot have large radii for their eddies, mixing even in strong winds is more laminar, and consequently rates of reaeration per area are small. Thus, one finds an apparent paradox that the thinner the water film, the more tendency it has to function anaerobically at night. (Odum, 1967)

2.4. Aquatic Systems with Very Short Anaerobic Exposures

Contiguous aerobic and anaerobic water layers are not completely separable from each other. Thus, short-term exposures to anaerobic conditions can be expected for communities living mainly under photoaerobic conditions, while organisms of photoanaerobic layers can also be exposed to temporary aerobic conditions. Changes in conditions may be due to vertical movements of the thermocline (Serruya, 1975), affecting populations living in the metalimnion, lower parts of the epilimnion, or upper parts of the hypolimnion. Vertical migration of the organisms, such as those of cyanobacteria, may have similar effects (Fogg and Walsby, 1971; Burgis *et al.*, 1973; Reynolds and Walsby, 1975; Walsby, 1975).

2.5. Sediments

Alternations in O_2 conditions described above for the water layers are commonly found in the mud bottoms (Baas Becking and Wood, 1955; Wood, 1965; Fenchel and Riedl, 1970; Fenchel, 1971; Sirenko, 1972; Burgis *et al.*, 1973; Jørgensen and Fenchel, 1974; Pfennig, 1975; Viner, 1975; Whitton and Sinclair, 1975; Cohen *et al.*, 1977c; Jørgensen, 1977; Jørgensen and Cohen, 1977). The extent of light penetration into the mud layer appears to be a

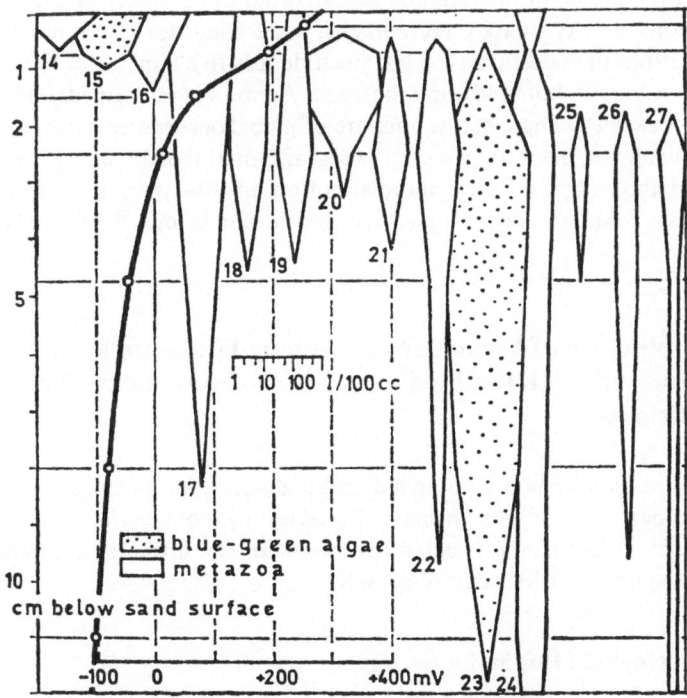

Figure 4. Vertical distribution of blue-green algae and metazoan species from Wrightsville (McCrary's) mudflats (North Carolina, U.S.A.), October. From Fenchel and Riedl (1970).

significant factor with respect to the upper surface layer (Fenchel and Straarup, 1971) and the undisturbed mud–water interface of shallow water bodies. Recently, Fenchel and Riedl (1970) (Fig. 4) described a "thiobiosis" with sulfide concentrations of 10 mM, with occasional exposure to light penetration possible. This habitat is found below an oxidizable surface layer of sand in the littoral zone of many seas. Alternation of O_2 tensions in the benthic habitats occurs in cycles similar to those of the water body (Fenchel and Riedl, 1970; Fenchel, 1971; Serruya et al., 1974). Phototactic or photophobic vertical movements of organisms through the benthic substrate expose them to a steep gradient of redox potential as in the water column (Castenholtz, 1969; Sournia, 1976).

2.6. Summary

Two extremes of light and O_2 conditions in aquatic systems have been described. Permanent photic aerobic conditions are found, as in the epilimnion of many lakes or the waters of the open seas. In contrast, stable sulfide-rich photoanaerobic conditions are found, as in the source waters of hot springs and the monimolimnion of meromictic lakes. In addition, there is a spectrum of combinations of oxygenated and anoxygenated conditions. The more frequent the alternations in conditions, the less possible it is to characterize the ecosystem as photoaerobic or photoanaerobic, and ecosystems with intermediate conditions must be recognized. These ecosystems are of global occurrence and of vast dimensions; for example, the mangroves, estuaries, and marine sediments ("thiobioses") mentioned above. In the future, more attention must be given to these important intermediate combinations of photoanaerobic conditions than has been done previously.

3. Predominance of Cyanobacteria among Phototrophs in Aquatic Systems with Alternating Photoaerobic–Photoanaerobic Conditions

In this section, data are compiled on the distribution of the different phototrophic types, i.e., photosynthetic bacteria, cyanobacteria, and eucaryotic phototrophs, found in the selected habitats with the different combinations of photoanaerobic conditions described in Section 2.

3.1. Phototrophs in Hot Sulfur Springs

The distribution of phototrophic organisms in neutral and alkaline hot sulfur springs of New Zealand, Europe, and the United States has been described

[Anagnostidis and Golubić, 1966; Castenholtz, 1969, 1973, 1976, 1977 (Section 2.1)]. There is a gradient in temperature, sulfide, O_2, pH, and other parameters downstream, and all accompanying changes in the flora can often be followed along one single stream (Fig. 1; Table I).

In the most sulfide-rich upstream waters of the New Zealand springs (Castenholtz, 1973, 1976) with temperatures of $60°-70°C$ and sulfide concentrations above 6 μM up to 2 mM, the filamentous photosynthetic gliding bacterium *Chloroflexus* (Castenholtz, 1973; Pierson and Castenholtz, 1974; Pfennig, 1977) occurs abundantly. It is the dominant phototroph of this system, forming thick, unibacterial, orange-green mats. The flora changes downstream along the temperature gradient: At $56°C$, where the sulfide concentration is still high ($\leqslant 2$ mM), *Oscillatoria amphigranulata* is dominant and forms a thick, orange-green layer. In these "*Oscillatoria* springs," *Chloroflexus* may still occur beneath the cyanobacterial layer, though apparently never alone. Occasionally at $52°-54°C$, *Synechococcus* sp. is also found. Further along the sulfide gradient, at $\leqslant 0.2$ μM ($53°-56°C$), *Mastigocladus laminosus* is the dominant species of the top layer, occasionally appearing with *O. amphigranulata. Chloroflexus* may occur beneath it. At somewhat higher temperatures ($58°-64°C$) with lower sulfide concentration ($\leqslant 5$ μM), *HTF Mastigocladus* is often dominant. *HTF* refers to the high-temperature form. It should be noted that the upper temperature limit of all strains decreases in the presence of sulfide.

A similar pattern is observed in Iceland (Castenholtz, 1976). Unibacterial *Chloroflexus* mats are observed upstream (15–60 μM sulfide, $62°-66°C$). In many of these springs, the sulfide concentration falls to zero at the point where the water cools to ca. $60°-64°C$; thus *HTF Mastigocladus* forms typical mats below that temperature. If the sulfide decrease occurs at ca. $54°C$, *M. laminosus* is dominant.

In the Yellowstone Park (U.S.A.) sulfur streams, a similar pattern of floral change is observed (Castenholtz, 1973) (Fig. 1; Table I). *Chloroflexus* dominates the upstream waters ($50°-65°C$) which contain 30–90 μM sulfide. It is often found below a mat of *Chlorobium* species. Downstream, at $50°C$, where sulfide is still present ($\leqslant 90$ μM), *Chlorobium* disappears. Here, *Chloroflexus* takes a dependent position below a cyanobacterium, *Spirulina labyrinthiformis. Synechococcus lividus* dominates at sulfide concentrations of less than 22 μM at $60°C$, while at $50°C$ and $\leqslant 6$ μM sulfide, *M. laminosus, S. lividus, S. minervae, Phormidium* sp., *Oscillatoria* sp., *Pseudoanabaena* sp., *S. labyrinthiformis*, and other cyanobacteria, as well as *Chromatium* sp., occasionally occur. *Oscillatoria terebriformis* is known to form mats at very low sulfide concentrations ($\leqslant 60$ μM) in springs in Oregon (U.S.A.).

Species are obviously excluded in waters where the temperature is above their maximum temperature limit; however, their presence in waters below this temperature is determined by other factors. The decline in photosynthetic

Table I. Instances of Cyanobacteria in Alternating Photoanaerobic to Photoaerobic Aquatic Systems

Cyanobacterial type	System and locality	Sulfide-S (mM)	Predominance among phototrophs
	Hot sulfur springs		
Oscillatoria amphigranulata[a]	New Zealand	≤2.0	100% or+ phot. bact.
Spirulina labyrinthiformis[b]	U.S.A. (Yellowstone)	≤0.3	100% or+ phot. bact.
Oscillatoria terebriformis[b]	U.S.A. (Oregon)	≤0.13	100% or+ phot. bact.
	Metalimnion of lakes		
O. agardhii Gom[c]	U.S.A., Canada, Germany		Forms plate
O. agardhii var. *isothrix* Skuja[d]	U.K., U.S.A.		Forms plate
O. rubescens DC ex Gom[e]	Austria, Switzerland, Sweden, Italy		Forms plate
O. limosa Ag. ex Born et Flah[f]	Czechoslovakia		Forms plate
O. prolifica Gom[g]	U.S.S.R.		Forms plate
O. utermoehliana Elenk.[h]	Germany		Forms plate
O. lauterbornii Schmidle	Canada		Forms plate
Spirulina laxa G-M Sm[i]	Canada		Forms plate
Oscillatoria redekei VanGoor[j]	Germany		Forms plate
Oscillatoria sp., *Microcystis* sp.[k]	Israel		Forms plate + phot. bact.
Anabaena planktonica Brunnth[l]	Germany		Forms plate

Hypolimnion of shallow lake			
Oscillatoria limnetica, O. salina, Microcoleus sp.[m]	Israel (Solar Lake)	≤1.5	Predominant or + phot. bact.
Polymictic, shallow equatorial lake			
Microcystic aeruginosa, M. flos-aquae, Anabaenopsis spp., Lyngbya spp., Aphanocapsa sp., Chroococcus sp.[n]	Uganda (Lake George)	Seldom and short-lived accumulation	80%
Shallow sea marshes			
Oscillatoria, Lyngbya, Microcoleus, Schizothrix, Anacystis spp.[o]	U.S.A. (Texas), Gulf of Mexico	Sulfide in bottom sediments	Forming mat
Littoral sea sediments			
Oscillatoria margaritifera, O. okenii, O. formosa, Merismopedia, Microcystis spp.[p]	World over	≤10–20 mM	Predominant

[a]Castenholtz (1976); [b]Castenholtz (1977); [c]Eberly (1964), Ohle (1964), Duthie and Carter (1970); [d]Lund (1959), Brook et al. (1971), Wohler and Hartmann (1973), Walsby and Klemer (1974); [e]Thomas (1949, 1950), Findenegg (1966), Ravera and Vollenweider (1968), Zimmerman (1969); [f]Stepanek and Chalupa (1958); [g]Gorlenko and Kusnezow (1972); [h]Elster (1966); [i]Schindler and Holmgren (1971); [j]Meffert (1975); Meffert and Krambeck (1977); [k]Serruya (1972); [l]Ohle (1964); [m]Cohen et al. (1977b); [n]Burgis et al. (1973), Ganf (1974c), Greenwood (1976); [o]Odum (1967); [p]Fenchel and Riedl (1970).

bacteria that occurs downstream cannot be unequivocally explained by changes in pH, sulfide, CO_2, and other factors. Castenholtz (1977) suggested that in Yellowstone Springs waters where sulfide is present, the turbulence, with consequent oxygenation and cooling of the water, may be responsible for the disappearance of *Chlorobium* and for the regression of *Chloroflexus* to a dependent position below a layer of cyanobacteria. O_2 production by the cyanobacteria may similarly exclude *Chlorobium* as an effective competitor. Temperature changes below the upper limit similarly cannot account for the cyanobacterial distribution (Castenholtz, 1976, 1977). The distribution of *HTF Mastigocladus* is not correlated with any chemical parameters except negatively with sulfide. It was observed to form mats in a great variety of chemical conditions in New Zealand hot springs with high ranges of pH, SiO_2, and Na^+, Cl^-, SO_4^{2-}, HCO_3^-, CO_3^{2-}, and F^- concentrations. There was similarly no correlation with other ions. Similar observations were made for *O. amphigranulata* (Castenholtz, 1977). A positive correlation of distribution with sulfide concentration is clearly evident for this cyanobacterium (Castenholtz, 1976, 1977) (Fig. 1). Hence, photosynthetic bacteria are the phototrophic dominants in the more purely photoanaerobic zone of the spring source, while downstream, cyanobacteria form a continuum dominating all intermediary sulfide concentrations to those of purely photoaerobic conditions.

No thermophilic eucaryotic algae have been shown to occur in the alkaline hot springs with or without sulfide. The eucaryotic alga *Cyanidium caldarum* dominates the acidic springs in the temperature range of 40°-57°C (Doemel and Brock, 1970; Brock, 1973b). Unfortunately, its relationship to sulfide has not been investigated.

3.2. Phototrophs of Stratified Lakes

The occurrence of photosynthetic bacteria in the anaerobic layers of stratified lakes (Section 2.2) has long been noted (Sorokin, 1965; Pfennig, 1967, 1975; Serruya, 1972; Caldwell, 1977; Cohen *et al.*, 1977b) (Fig. 8). Thus Chromatiaceae or Chlorobiaceae or both have been shown to form plates in the monimolimnion of meromictic lakes, hypolimnion and metalimnion of monomictic and dimictic lakes, as well as on muddy bottom layers penetrable by light. Recently, Biebl and Pfennig (1977) compiled the published data on phototrophic bacteria from the viewpoint of primary production in lakes (see also Table 1 in Cohen *et al.*, 1977b); their survey reveals that photosynthetic sulfur bacteria can contribute 3-90% of the total production. For the Fayetteville Green Lake (N.Y., U.S.A.), it was shown (Culver and Brunskill, 1969) that although *Chlorobium* grows at 17-20 m below the surface, where the O_2 content is low, it is grazed by various zooplankton species predominant in the upper layers. Thus, the biomass produced is not only recycled within the hypolimnion but may enter the food chain of the entire lake.

While photosynthetic bacteria dominate the anaerobic habitats, eucaryotic algae mainly dominate the photoaerobic epilimnion of lakes; their importance in the primary production of these water bodies is well known (Berman and Pollingher, 1974; Wetzel, 1975). While the distribution of photosynthetic bacteria and eucaryotic algae is restricted to zones with photoanaerobic or photoaerobic conditions, respectively, the cyanobacteria occur in the epilimnion (Walsby, 1970; Findenegg, 1971; Wetzel, 1975; Reynolds and Walsby, 1975) as well as in the meta- and hypolimnion (Table I) (Utermöhl, 1925; Juday, 1934; Findenegg, 1966; Meffert, 1975; Reynolds and Walsby, 1975; Whitton and Sinclair, 1975; Reynolds, 1976; Caldwell, 1977; Meffert and Krambeck, 1977).

The cyanobacteria component in the phytoplankton becomes increasingly important the more eutrophic the lake (Bozniak and Kennedy, 1968; Walsby, 1970; Lund, 1973; Reynolds, 1973; Gorham et al., 1974; Reynolds and Walsby, 1975; Wetzel, 1975; Whitton and Sinclair, 1975; Caldwell, 1977; Robarts and Southall, 1977). This phenomenon has economic implications where sources of water supply are concerned, as blooms of cyanobacteria often cause deterioration of the quality of the water (Walsby, 1970; Shilo, 1972; Reynolds and Walsby, 1975). They introduce odor and taste into drinking water, clog irrigation pipes, spoil recreation areas, introduce toxins to livestock feed, and may bring about collapse of the O_2 tension balance of the waters. The high concentration of organic material in eutrophic water bodies has been related to this prevalence of cyanobacteria. However, in view of the restricted heterotrophic capacity of cyanobacteria (Smith and Hoare, 1977; Stanier and Cohen-Bazire, 1977), this assumption is questionable (Reynolds and Walsby, 1975). As mentioned above (Sections 2.2 and 2.3), the O_2 budget in the eutrophic lake ecosystem, even in the epilimnion, is given to frequent alternations between aerobic and anaerobic conditions, especially marked at night. The possibility that these alternations may contribute to cyanobacterial proliferation in eutrophic lakes has been proposed (Walsby, 1970; Fogg et al., 1973; Reynolds and Walsby, 1975). It may also be of significance that the seasonal growth of *Microcystis* in some temperate lakes coincides with the onset of reducing conditions in the hypolimnion (Reynolds and Walsby, 1975; Reynolds and Rogers, 1976).

The correlation of frequent alternations in O_2 tension with abundance of cyanobacteria is particularly well illustrated in the equatorial lakes; the more frequent the stratification cycle, the more prominent the cyanobacteria populations. Thus, in the polymictic lakes of the equatorial region, the dominating presence of cyanobacteria has long been recorded (Fritsch, 1907; Rich, 1932; Singh, 1955; Damas, 1964; Iltis, 1971; Fogg and Walsby, 1971; Talling et al., 1973; Reynolds and Walsby, 1975; Whitton and Sinclair, 1975; Caldwell, 1977). A detailed study has been carried out in Lake George of both the stratification cycle (Section 2.2; Ganf and Viner, 1973; Viner and Smith, 1973; Ganf, 1974a,b,c; Ganf and Horne, 1975) and biological parameters (Ganf and Viner, 1973; Burgis et al., 1973; Ganf, 1974a,b,c; Ganf and Horne, 1975; Greenwood,

1976). The daily stratification pattern of Lake George may cause it to be regarded as having a 24-hr physiological cycle rather than a seasonal one (Fig. 3a–d). The marked diurnal changes in solar radiation, temperature, pH, and O_2 have a predominant effect on the flora and fauna. Other changes throughout the year are almost negligible in this tropical climate and within this catchment area. The phytoplankton biomass is extremely high, 30–48 g C/m^2 (the depth is only 2.4 m), and forms 95% of the total biomass; of this, 80% is cyanobacteria (Table I). At dawn, when the waters are isothermal and mixed, the phytoplankton is uniformly distributed. With the onset of stratification, it tends to sink. In the evening, the highest algal concentration is found below the euphotic zone. Redistribution occurs after sunset when stratification begins to break down (Fig. 3d).

The depth and turbulence cycle of Lake George results in an accumulation of settled but viable phytoplankton in the superficial sediments of the lake which are anoxic for the most part. When these sediments are stirred, the algae are resuspended. Samples taken in the water column on such occasions show no significant increase of chlorophytes and diatoms, but the number of cyanobacteria increases 2–7 times. Although their relative viability decreases noticeably with depth in the sediments, these cyanobacteria of the mud (upper 10 cm) form an integral part of the water-column metabolism (Burgis *et al.*, 1973; Ganf, 1974a,b; Greenwood, 1976).

A striking feature of Lake George is its superficial stability, inferred from the stability of the phytoplankton biomass over the year. Yet the occurrence of such phenomena as sudden fish kills (Ganf and Viner, 1973) reveals that the lake is easily upset by slight ecological changes. Greenwood (1976) suggested that the relative long-term stability of the equatorial climate may have allowed the biota to evolve a delicate steady state with respect to the short-term environmental changes. Not only is the total biomass stable, but also the species composition— a small number of cyanobacteria species—does not undergo succession (Table I). The short stratification cycle of Lake George has a time span (24 hr) which is less than the generation time of the majority of the phytoplankton, insufficient time for the species composition to change. Only species that can resist the whole range of daily alternations of Lake George will survive; these are large colonial types of cyanobacteria: *Microcystis* and *Aphanocapsa* predominating, with filamentous forms like *Anabaenopsis* sp., *Lyngbya* sp., *Aphanizomenon* sp., and *Anabaena* sp. present (Burgis *et al.*, 1973; Ganf, 1974c; Greenwood, 1976).

In the very special cyanobacterial ecology, Lake George should be placed high in the continuum of lake types ranging from versatile phytoplankton communities to "cyanobacterial lakes." These same lakes can then be placed in a continuum of increasing frequencies of marked changes in O_2 tension, pH, and temperature.

3.3 Phototrophs of Shallow Water Systems

As already mentioned (Section 2.3.), similar marked diurnal alternations in O_2 tension, pH, and temperature characterize the shallow waters of marshes, mangroves, and rice paddies. The predominance of cyanobacteria in these habitats has been recognized for a long time (Abeliovich, 1967; Sirenko et al., 1969; Topachevskii et al., 1969; Fogg and Walsby, 1971; Fogg et al., 1973; Brock, 1973a,b; Reynolds and Walsby, 1975; Whitton and Sinclair, 1975). The sea marshes described by Odum (1967) around the Gulf of Mexico (see Section 2.3.) include very different habitats with respect to depths, salinities, organic materials, and chemicals. As this author notes, the thinner the water body, the greater the fluctuations in parameters governed by the diurnal cycle, i.e., O_2, temperature, and pH. This shallow ecosystem (less than 10 cm deep) is characterized by a specialized flora (and fauna)—cyanobacteria of several genera: *Lyngbya, Oscillatoria, Microcoleus, Schizothrix,* and *Anacystis* (Table I). These species form an interwoven mat of filaments, which may become as much as 1 cm thick, with a very high chlorophyll concentration (300–600 mg chlorophyll a/m^2). This highly compressed, shallow water body is characterized by sulfides at the bottom with a vertical gradient in redox potential and other parameters; the redox potential gradient may produce a movement of negative ions toward the positive upper surface. It is suggested that nutrients such as phosphate and nitrate move upward; thus, some of the energy created by the voltage difference may be used to pump nutrients (Odum, 1967). When consumers are restricted, the deposition of organic matter is high, favoring subsequent oil formation (Odum, 1967).

3.4. Phototrophs of Sediments

In the sediment system proper (Section 2.5), proliferation of photosynthetic bacteria and cyanobacteria is known to take place (Wood, 1965; Pfennig, 1967, 1975; Fenchel and Straarup, 1971; Sirenko, 1972; Fogg et al., 1973; Golubić, 1973; Doemel and Brock, 1974; Jørgensen and Fenchel, 1974; Blackburn et al., 1975; Viner, 1975; Whitton and Sinclair, 1975; Sournia, 1976; Cohen et al., 1977c; Jørgensen and Cohen, 1977; Krumbein and Cohen, 1977; Potts and Whitton, 1977; Potts et al., 1977). Recently, Fenchel and Riedl (1970) described (Section 2.5; Fig. 4; Table I) the "thiobioses" found below oxidizable littoral sands of most seas:

> Typically motile filamentous types, such as *Oscillatoria margantifera, O. okenii, O. formosa,* etc., predominate in density and diversity, followed by a few Chroococcales. . . . they are more diverse and dense and increasingly conquer the depths of the environment when grain size and sorting decrease, porosity and permeability drop, and the redox-potential-discontinuity layer becomes distinct and moves closest to the surface.

Several species reach a population density of 5000 colonies or filaments/100 cm^2. It should be added that these "thiobioses" are thought to play a major role in the energy cycles of marine ecosystems and to represent a possible first stage in petroleum production.

3.5. Phototrophs of the Open Sea

In the completely photoaerobic zones of most seas, eucaryotic algae are the main phytoplankton forms (Strickland, 1965; Wood, 1965; Steemann-Nielsen, 1975). Few isolates of marine photosynthetic bacteria are known (Wood, 1965). Similarly, cyanobacteria are rare in the open sea; some instances of *Trichodesmium* blooms have been encountered in oligotrophic stratified seas of the equatorial region (Wood, 1965). These blooms have been related to the cyanobacterial capacity to fix N_2 (Dugdale *et al.*, 1961; Taylor *et al.*, 1973; Carpenter, 1973; Carpenter and Price, 1977).

3.6. Summary

Taking into account only the alternation in O_2 tension and sulfide concentrations in the photic systems described, it is possible to consider that a gap exists at the ecological level between the photosynthetic bacteria and the eucaryotic algae; the former dominate the most stable photoanaerobic situations, and the latter, the most stable photoaerobic ones. No case has been recorded of photosynthetic bacteria proliferating photosynthetically in photoaerobic domains, and no clear-cut evidence exists for eucaryotic algae growing in a sulfide-rich photoanaerobic ecosystem (Kessler, 1974). Furthermore, both these phototrophs avoid ecosystems with marked and rapid alternating anaerobic and aerobic conditions. The cyanobacteria seem to bridge this apparent gap in dominating the ecosystems intermediate with respect to O_2 and sulfide tension.

In view of the inadequacy of available data, this generalization is clearly over-simplified. Changes in parameters other than O_2 and sulfide, i.e., light, pH, temperature, bound nitrogen, other organic and inorganic nutrient concentrations, etc., are thought to have a selective role in determining the cyanobacterial population size and composition (Walsby, 1970; Reynolds and Walsby, 1975; Whitton and Sinclair, 1975); thus, the relative importance of these factors should be tested.

It should be emphasized, nevertheless, that predominance of cyanobacteria in microaerophilic and/or sulfide-containing habitats has been recognized for a long time (Hinze, 1903; Nakamura, 1937, 1938; Huber-Pestalozzi, 1938; Czurda, 1940/1941; Fritsch, 1945; Schuster, 1949; Allen, 1952; Baas Becking and Wood,

1955; Setlike, 1957; Schwabe, 1960; Singh, 1961; Wood, 1965; Knobloch, 1966b; Stewart and Pearson, 1970; Fogg et al., 1973; Brock, 1973a, b; Serruya et al., 1974; Whitton and Sinclair, 1975). According to Wundsch (1940) and Huber-Pestalozzi (1938), Oscillatoria aghardhii is an indicator organism for the presence of free hydrogen sulfide. Brock (1973a,b) has suggested that the ability of cyanobacteria to grow under low O_2 concentrations may be one of the most important factors controlling their distribution.

The proposed ranking of the different photosynthesizing groups along a continuum with different oxygenation and sulfide conditions calls for analysis at the physiological level. Analysis of the phototrophic physiology of the three groups can provide information toward validation of this proposed ecological model.

4. Phototrophic Physiology of Photosynthetic Organisms

As with the ecological level, a wide gap is observed between photosynthetic bacteria and eucaryotic algae at the physiological level, and cyanobacteria seem to fit into this gap. The question arises as to the extent that one can relate the proposed ecological ranking to observations at the physiological level.

4.1. Photosynthetic Bacteria

In general, the photosynthetic system of photosynthetic bacteria (reviews in Pfennig, 1967, 1977; Frenkel, 1970; Gest, 1972; Parson 1974; Gromet-Elhanan, 1977) consists of one photosystem by which light energy absorbed by chlorophyll initiates an electron flow through electron and/or hydrogen carriers where a gradient of redox potential exists. The electrons finally return to the oxidized chlorophyll, and a cyclic electron flow is established. Through this light-initiated electron flow down the redox potential gradient, energy is conserved in forms utilizable by the cell, i.e., high-energy is conserved in forms utilizable by the cell, i.e., high-energy bonds of ATP and electric and osmotic gradients. As the electron flow is cyclic, there is no net generation of reducing power, and the power required for the cell must be derived from an exogenous source: organic substrates or highly reducing inorganic substances such as H_2, sulfide, or thiosulfate. The way the cells' reducing power is derived from these external sources is usually indirectly linked, if at all, with the photosynthetic system (Gest, 1972).

The phototrophic growth physiology of photosynthetic bacteria is anaerobic. This is true both for obligately and also for facultatively anaerobic photosynthetic bacteria (Cohen-Bazire et al., 1957; Lascelles, 1959; Pfennig, 1967, 1977).

4.2. Eucaryotic Algae and Plants

The photosynthetic system of eucaryotic phototrophs (reviews in Trebst, 1974; Govindjee and Govindjee, 1975; Jagendorf, 1975; Avron, 1977) is based on two photosystems (I and II) operating in series. Light absorbed by chlorophylls of these systems maintains an electron flow, in this case an open flow from the water-electron donor to the NADP-electron acceptor. Thus, energy is conserved at the same time in NADPH, ATP, and chemical and electrical gradients. Hence, in contrast to the bacterial photosynthetic system, net reducing power (NADPH) is regenerated in the eucaryotic-type photosynthesis through photosplitting of water with O_2 evolution. The photosynthetic pattern of bacteria is thus anoxygenic, whereas that of the eucaryotic phototrophs is oxygenic.

Nevertheless, already in the early sixties, it was shown that the oxygenic photosynthesis of eucaryotic algae can be turned into a functional, bacterial-type anoxygenic photosynthesis. If photosystem II is inhibited by its specific inhibitor, DCMU [3-(3,4-dichlorophenyl)-1,1-dimethylurea], or if it is not excited when light is provided in wavelengths absorbed only by photosystem I, a cyclic electron flow is maintained around photosystem I with energy-conserving patterns similar to those in photosynthetic bacteria. Arnon et al. (1961) suggested that this situation may represent an intermediate stage between plant oxygenic and bacterial anoxygenic photosynthesis. Moreover, under these conditions, the eucaryotic phototrophs photoassimilate organic substances in a similar way to bacteria (Simonis and Urbach, 1973; Kessler, 1974). Gaffron (1944; and see Kessler, 1974) showed that after adaptation to H_2, these cells can photoreduce CO_2 with H_2 serving as an electron donor as it does in many photosynthetic bacteria.

In contrast to the photosynthetic bacteria, however, eucaryotic phototrophs use sulfide as an electron donor very inefficiently, if at all (Knobloch, 1966a, b, 1969; Kessler, 1974). Moreover, eucaryotic algae differ markedly from photosynthetic bacteria in that they cannot grow photoautotrophically in anaerobic conditions, even with H_2 as an electron donor (Gaffron, 1944; Lewin, 1950; Nührenberg et al., 1968; Kessler, 1974). Although aerobic photoheterotrophic growth occurs in eucaryotic algae (Simonis and Urbach, 1973; Kessler, 1974), no unequivocal evidence exists for their anaerobic photoheterotrophic growth or for anaerobic heterotrophism (Gibbs, 1962; Simonis and Urbach, 1973; Kessler, 1974). Since anoxygenic photosynthesis as well as heterotrophic capacity is not in itself sufficient to allow anaerobic growth, the eucaryotic phototrophs appear to be obligate aerobes.

4.3. Cyanobacteria

The photosynthetic system of cyanobacteria is very similar to that of eucaryotic phototrophs; it is a eucaryotic type of oxygenic photosynthetic system

(Krogmann, 1973, 1977) in which photosystem II can be inhibited or inactivated, leaving only cyclic electron flow around photosystem I. Under these conditions, organic substances can be photoassimilated and support aerobic growth (Smith *et al.*, 1967; Rippka, 1972; Smith and Hoare, 1977; Raboy and Padan, 1978), and H_2 can serve as an electron donor to photoassimilate CO_2 or be evolved (Benemann and Wear, 1974; Weissman and Benemann, 1977; Bothe *et al.*, 1977; Tel-Or *et al.*, 1977).

The photosynthetic system of cyanobacteria differs from that of eucaryotic algae; the cyanobacteria have been shown to be capable of using sulfide as an electron donor, in a process similar to that of photosynthetic bacteria. In *O. limnetica*, a cyanobacterium isolated from the sulfide-rich layers of the Solar Lake (Elat, Israel), Cohen *et al.* (1975a) found oxygenic photosynthesis to be inhibited by DCMU (Fig. 5a). When sulfide is added to the oxygenic system, immediate inhibition of CO_2 photoassimilation is obtained; this inhibition is found with very low concentrations of sulfide (0.1–0.2 mM) (Cohen *et al.*, 1975a; Oren and Padan, 1978) (Fig. 5a). However, if exposure to light is con-

Figure 5. Anoxygenic photosynthesis in *Oscillatoria limnetica*. Data compiled from Garlick *et al.* (1977), Oren and Padan (1978), and Oren and Padan (unpublished data). The rate of CO_2 photoassimilation was determined in photoautotrophic aerobically grown cells resuspended in fresh growth medium with the indicated additions and incubated at 35°C and pH 6.8 in the light (60-W tungsten lamps, 2×10^4 ergs/sec per cm^2). a: Effect of Induction. - - -, No additions, or with CM (chloramphenicol) added at time zero. · · · ·, DCMU added at time zero. ———, Na$_2$S with or without DCMU added at 1 hr incubation time. —·—·—, Na$_2$S and CM with or without DCMU added at 1 hr incubation time. Concentrations: CM, 10 μg/ml; DCMU, 5 μM; Na$_2$S, 3.5 mM. b: Effect of sulfide concentration. The rate of CO_2 photoassimilation at 703-nm light in the presence of DCMU was measured at various Na$_2$S concentrations. (X), *Lyngbya* (7104); ($\bullet$), *Aphanothece halophytica* (7418); ($\circ$), *O. limnetica*. c: Effect of pH. The rate of CO_2 photoassimilation of induced cells was measured at different medium pHs in the presence of 4 mM Na$_2$S and 5 μM DCMU. The incubation medium contained 5mM Hepes (*N*-2-hydroxyethyl-piperazine-*N'*-ethanesulfonic acid) and growth medium lacking phosphate.

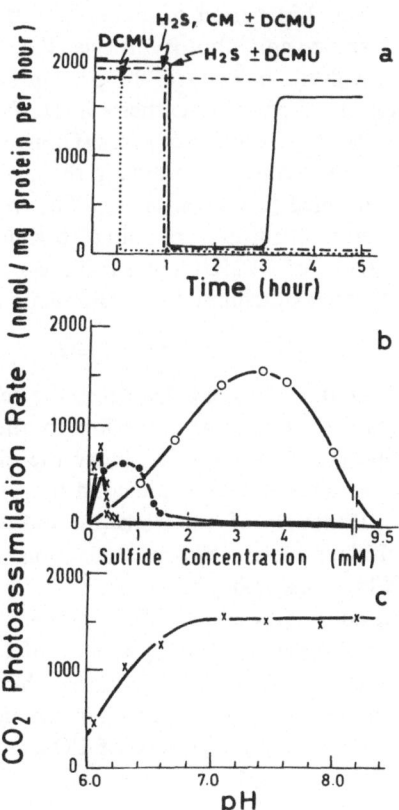

tinued for about 2 hr in the presence of a high sulfide concentration (3mM), the photoassimilation reappears, but it is now insensitive to DCMU (Cohen *et al.*, 1975a; Oren *et al.*, 1977; Oren and Padan, 1978) (Fig. 5a). The reappearing photoassimilation is therefore anoxygenic, independent of photosystem II, and driven by photosystem I, with sulfide as an electron donor. Both light and sulfide are absolute requirements for this anoxygenic photosynthesis.

The possible participation of photosystem II in the reaction was further excluded by results of experiments in which photosystem II was not activated (rather than inhibited) in the presence of far-red light (>673 nm). Under this light condition, oxygenic photosynthesis, requiring operation of both photosystems, was drastically inhibited ("red drop"), whereas anoxygenic photosynthesis in the presence of sulfide was fully operative (Oren *et al.*, 1977). Furthermore, if both photosystems were operative and contributed to the reaction, an enhancement in quantum yield of the photoassimilation would be predicted with respect to the yield obtained with one photosystem in operation. However, the "enhancement phenomenon" was observed only with oxygenic photosynthesis, while none was observed with the anoxygenic process (Oren *et al.*, 1977).

The 2-hr period of incubation in the presence of sulfide and light, required for anoxygenic photosynthesis (Fig. 5a), may indicate that an induction process is involved. Indeed, in the presence of chloramphenicol, which is a specific inhibitor of protein synthesis in *O. limnetica* and in bacteria, anoxygenic photosynthesis does not take place (Oren and Padan, 1978).

The fate of the sulfide in the anoxygenic photosynthesis of *O. limnetica* was determined (Cohen *et al.*, 1975b) from the distribution of the radioactive label in different sulfur-containing fractions of a culture incubated with $Na_2{}^{35}S$ in the presence of DCMU: The sulfide is oxidized to elemental sulfur according to the following stoichiometric relationship (see also Cohen *et al.*, 1975a):

$$2 H_2S + CO_2 \longrightarrow HCHO + 2 S^0 + H_2O$$

Elemental sulfur expelled from the cells was observed in the medium as refractile granules, sometimes adhering to the cyanobacterial filaments (Fig. 6a). The appearance of the cells under the electron microscope is the same under both anoxygenic and oxygenic conditions (Fig. 6b).

Donation of electrons by sulfide, resulting in evolution of H_2, occurs in sulfide-induced cells in a reaction which is light and sulfide dependent and DCMU insensitive. The reaction is accelerated by FCCP (carboylcyanide P-trifluoromethoxyphenylhydrazone), the uncoupler of electron-transport-coupled phosphorylation. The presence of CO_2 inhibits this photosystem-I-driven H_2 evolution, while FCCP relieves the inhibition. Acceleration by FCCP implies that hydrogenase rather than nitrogenase is involved in the reaction. The light requirement and effects of CO_2 and uncoupler in this reaction indicate that

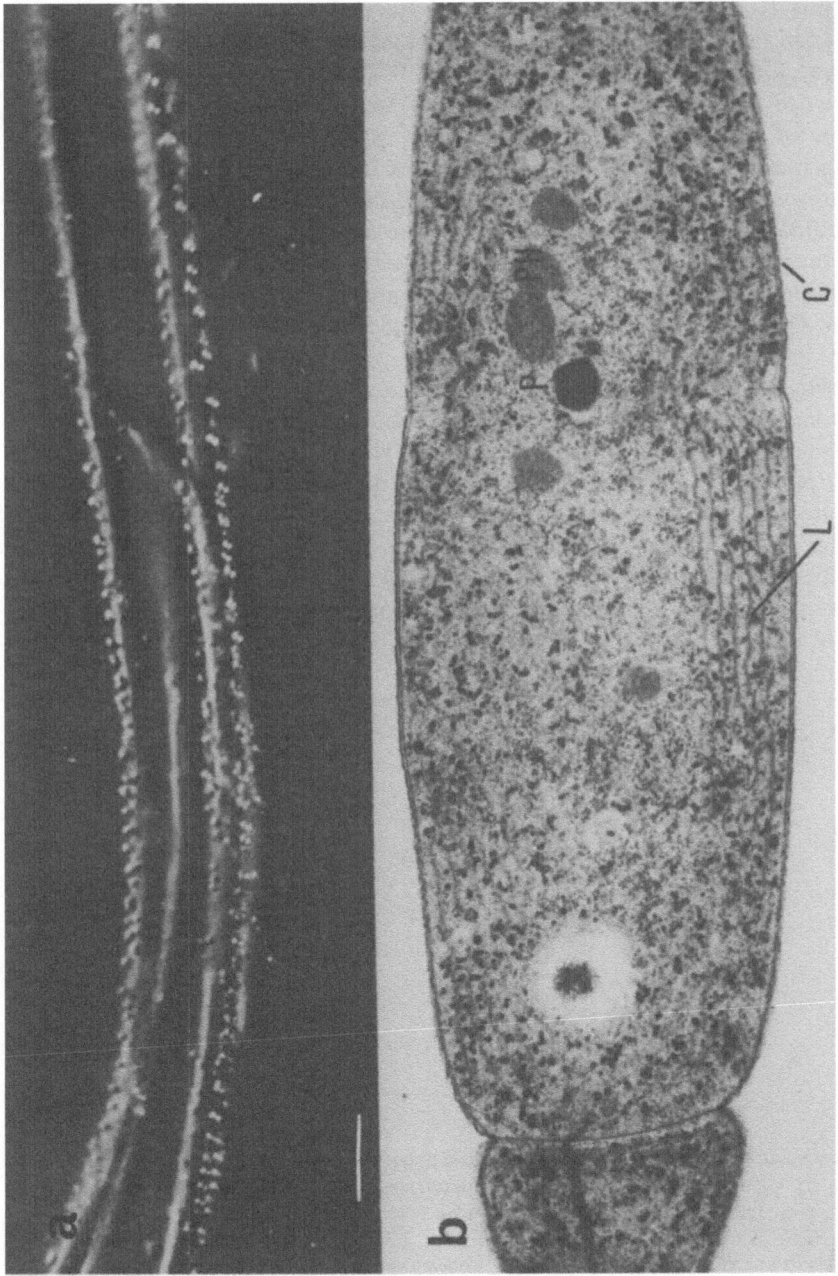

Figure 6. *Oscillatoria limnetica* filaments grown in anoxygenic conditions in the presence of sulfide (Oren and Padan, 1978). a: Nomarski interference contrast; the bar represents 10 μm. Courtesy of M. Kessel. b: Electron micrograph; the bar represents 1 μm. C, Cell wall; L, photosynthetic lamellae with α granules; P, polyphosphate granule; PH, polyhedral body. Courtesy of A. Oren and S. C. Holt.

sulfide donates electrons directly to the photosynthetic system, with protons serving as an electron acceptor (Belkin and Padan, unpublished data).

As yet, a cell-free preparation oxidizing sulfide has not been obtained from *O. limnetica* cells. In bacteria, cytochromes have been shown to be primary acceptors of electrons from various sulfur-containing electron donors (Kusai and Yamanaka, 1973a,b; Trüper, 1973; Fischer and Trüper, 1977). Two soluble cytochromes could be detected (Oren, unpublished data) in aqueous extracts of both sulfide-induced and noninduced *O. limnetica* cells, with amounts of the cytochromes similar in both preparations. These results are comparable with other cyanobacterial systems (Susor and Krogmann, 1966; Holton and Myers, 1967a,b; Ogawa and Vernon, 1971; Cramer and Whitmarsh, 1977). The first is a high-potential cytochrome, which is reduced by ascorbate as well as sulfide and comprises 81% of the total soluble cytochrome. The second is a low-potential cytochrome which is reduced by dithionite but not by ascorbate or sulfide (even under N_2 in the presence of up to 50 mM Na_2S at pH 8 or 7.3).

The pattern in which *O. limnetica* utilizes light energy in its two types of photoautotrophic growth patterns was investigated by Oren *et al.* (1977). In *O. limnetica*, light is absorbed by chlorophyll a, phycocyanin, and carotenoids. The cell concentrations of these pigments are identical under both physiological conditions. Quantum-yield spectra (Fig. 7) of oxygenic and anoxygenic photo-

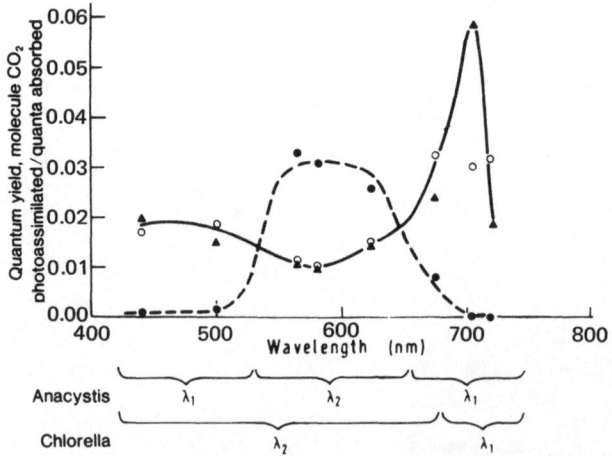

Figure 7. Quantum-yield spectra of oxygenic and anoxygenic photosynthesis of *Oscillatoria limnetica*. Oxygenic photosynthesis (•) was determined in cells grown and incubated in sulfide-free medium. Anoxygenic photosynthesis was determined in cells grown in the presence of Na_2S and incubated in the presence of 3.5 mM Na_2S (o) or in the presence of 3.5 mM Na_2S and 5 μM DCMU (▲). The λ_1 and λ_2 spectral regions (wavelengths preferentially exciting photosystems I and II, respectively), defined by Jones and Myers (1964) for *Anacystis* and Govindjee *et al.* (1968) for *Chlorella*, are shown. From Oren *et al.* (1977).

synthesis reveal that, whereas the entire absorbed spectrum is utilized in the anoxygenic photosynthesis, only a narrow sector is utilized in the oxygenic type. A marked drop in efficiency at both blue and red wavelengths is observed with the oxygenic photosynthesis, and even the relatively inefficient anoxygenic photosynthesis, at 560–650 nm, is far more efficient. This limited quantum utilization in oxygenic photosynthesis of *O. limnetica* is markedly different from that of eucaryotic algae and plants containing chlorophyll b. In these, almost the entire absorbed spectrum is utilized in the oxygenic photosynthesis, excluding only the far-"red drop" (Govindjee *et al.*, 1968). Hence, it is suggested that the pattern of quantum utilization in the photosynthetic system of *O. limnetica* is oriented less toward oxygenic photosynthesis than that in higher eucaryotic phototrophs. It nevertheless allows efficient utilization of the absorbed spectrum in anoxygenic photosynthesis.

Once induced, *O. limnetica* can readily grow photoautotrophically anaerobically. This was demonstrated by Oren and Padan (1978) using a growth system with a N_2 gas phase in which both the sulfide concentration and pH were kept constant. The anaerobic photoautotrophic growth rate (doubling time, 2 days) is similar to the aerobic photoautotrophic rate. Once sulfide is excluded, the anaerobically grown *O. limnetica* immediately shifts to photoaerobic growth with oxygenic photosynthesis, and the anoxygenic photosynthetic capacity is diluted out by growth. Hence, *O. limnetica* possesses facultatively anoxygenic photosynthetic capacity and is a facultatively anaerobic photoautotroph which readily shifts between aerobic photoautotrophic and anaerobic photoautotrophic growth patterns. This phototrophic physiology is unique in view of the restrictions of photosynthetic bacteria and eucaryotic phototrophs to either aerobic or anaerobic growth conditions.

4.4. Summary

The facultatively anoxygenic phototrophic physiological pattern of *O. limnetica*, utilizing sulfide with the entire absorbed spectrum, is thought to interlink the apparent strictly aerobic phototrophic physiology of the eucaryotic algae and the strictly phototrophic anaerobic one of the photosynthetic bacteria. Thus, the continuity observed between the different phototrophs at the physiological level parallels the ecological pattern exhibited by them in the photoanaerobic ecosystems.

The dependence of photosynthetic bacteria on exogenous reducing substances for their generation of reducing power is compatible with their occurrence in photoanaerobic ecosystems in which such substances prevail. Accordingly, the wide occurrence of eucaryotic phototrophs in photoaerobic domains corresponds with their capacity for oxygenic photosynthesis. The prevalence of cyanobacteria in ecosystems which undergo alternations in O_2 tension and sulfide concentration corresponds with a capacity for facultatively anoxygenic photosynthesis such as that exhibited by *O. limnetica* in the presence of sulfide. However, the causes for the restriction of eucaryotic algae and photosynthetic bacteria to respective aerobic or anaerobic growth conditions while *O. limnetica* grows under both conditions are not self-apparent.

The incapability of eucaryotic algae to photoassimilate CO_2 and grow in the presence of sulfide may be explained by the toxicity of sulfide either to the photosynthetic system or to a different system (Evans, 1967; Izawa and Good, 1972; Good and Izawa, 1973). Nevertheless, the reasons these algae do not grow anaerobically heterotrophically or phototrophically in the presence of H_2 have not been determined.

Anaerobic growth physiology includes at least two indispensable requirements. The first prerequisite is the noninvolvement of O_2 in any essential enzymatic reaction. Eucaryotic phototrophs possess enzymes which require O_2, such as those involved in fatty acid synthesis, sterol synthesis, and phenol oxidation (Goad and Goodwin, 1972; Wood, 1974; Hayaishi, 1974; Hamberg *et al.*, 1974). However, it is not yet certain which of these enzymes are not replaceable by alternative pathways which may operate anaerobically. Such O_2-indispensable pathways exist in *Saccharomyces*. These organisms, however, are grown anaerobically if supplied exogenously with essential products such as sterols (Andreasen and Stier, 1953). Strangely enough, no such anaerobic growth experiments have been recorded with eucaryotic phototrophs.

The second prerequisite for anaerobic growth is a capacity to function under reducing conditions, reaching -200 to -300 mV due to high concentrations of sulfide or H_2 (Baas Becking and Wood, 1955; Fenchel and Riedl, 1970; Gest, 1972; Pfennig, 1975; Cohen *et al.*, 1977a,b). The second law of thermodynamics requires that electron and ion movements across the cell membrane will equalize the redox potential difference between the internal and the external environmental medium unless continual work is being done to reverse this tendency. The organic matter of the living system must remain in its reduced condition. Hence, the electron and hydrogen carriers of the cell must be poised at a corresponding low redox potential. Thus, growth capability in a particular environmental redox potential will be determined by both the particular internal redox potential and the capacity to maintain it under the respective conditions. When the system is in a reduced environment, it has the reverse problem—to pump reducing species out into the environment. H_2 evolution is an example of such an adaptive process in photosynthetic bacteria (Gest, 1972). In this context, it is possible that the incapability of eucaryotic algae to grow anaerobically stems from their inability to maintain their redox potential under anaerobic conditions, resulting in stringent dependence on photoaerobic conditions (Gibbs, 1962).

Both these essential requirements for anaerobic growth are met by photosynthetic bacteria growing photoanaerobically; these do not need O_2 in any metabolic reaction and endure low external redox potentials. Furthermore, for obligately anaerobic photosynthetic bacteria, as in other obligate anaerobes, O_2 is toxic to several metabolic steps (McCord et al., 1971). The negative control exerted by O_2 on the photosynthetic system of facultatively anaerobic bacteria is also a well-known phenomenon (Cohen-Bazire et al., 1957; Lascelles, 1959). The photosynthetic system of both these facultative and obligate anaerobes appears to operate most efficiently when poised at a negative redox potential (Frenkel, 1970). This is understandable in this cyclic system operating without change in net oxido-reduction potential, functioning in a low-redox-potential milieu.

O. limnetica growing anaerobically in the presence of high sulfide concentrations (3 mM at pH 6.8 and 12 mM at pH 8) endures low levels of redox potential. Recently, Oren, Tietz, and Padan (1978, unpublished data) showed that this cyanobacterium contains only unsaturated and monounsaturated fatty acids, whether grown aerobically or anaerobically. Its mechanism of fatty acid desaturation does not require O_2. This part of fatty acid metabolism of O. limnetica differs markedly from that of many filamentous cyanobacteria, in which a high content of polyunsaturated fatty acids has been found (Kenyon et al., 1972) along with an O_2 requiring desaturation mechanism (Bloch et al., 1966).

In view of the restrictions of eucaryotic algae or photosynthetic bacteria to exclusively photoaerobic or photoanaerobic conditions, respectively, a cyanobacterium with an O. limnetica-type facultatively anoxygenic phototrophy should have a selective advantage in an ecosystem which alternates between aerobic and anaerobic conditions. The more frequent and marked the alternations, the more prominent the cyanobacteria component with respect to the other phototrophic organisms.

5. Facultatively Anoxygenic Photosynthetic Cyanobacteria in Nature

5.1. Spread of Anoxygenic Photosynthesis among Cyanobacteria

We must now consider whether O. limnetica is exceptional among the cyanobacteria and whether its physiological pattern may be interpreted, in consequence, as secondary specialization toward anaerobic photosynthesis. Garlick et al. (1977) (Table II) tested various cyanobacteria for the property of anoxygenic photosynthesis, including strains provided by R. Y. Stanier as well as many new isolates. Castenholtz (1976, 1977) (Table II) tested cyanobacteria of alkaline hot sulfur springs of New Zealand and Yellowstone (U.S.A.).

All the strains were capable of oxygenic photosynthesis inhibited by DCMU

Table II. Sulfide-Dependent Photosystem-I-Driven Anoxygenic Photoassimilation of CO_2 among Cyanobacteria

Cyanobacteria	Maximal measured rate (nmol of CO_2/mg protein per hr)	Sulfide concentration (mM)			Inhibiting oxygenic photosynthesis[k]
		At which anoxygenic photosynthesis took place			
		Detected	Maximal	Tested	
Oscillatoria limnetica[a,c]	1600	1-8.5	3.5		0.1
O. salina[a,c,l]	2270	0.45-0.9	0.45		
Oscillatoria (6407)[a,d]	12	0.77	0.77		
Oscillatoria sp.[a,g,l]	146	0.83	0.83		
O. amphigranulata[b,e]	n.d.	0.3-2	1-2		<0.6
O. terebriformis[b,f]	n.d.	0.3-1	0.8		
Spirulina labyrinthiformis[b,f]	n.d.	0.2-1.8	0.6		<0.6
Microcoleus sp.[a,c,l]	500	0.45-0.9	0.9		
Phormidium sp.[a,h,l]	540	0.5-0.8	0.5		
Phormidium sp.[a,i,l]	158	0.35-1.7	0.35		
Phormidium sp.[a,i,l]	1040	0.68	0.68		
Lyngbya (7004)[a,d]	3750	0.05-4.1	0.28		
Lyngbya (7104)[a,d]	751	0.08-0.4	0.12		
Aphanothece halophytica (7418)[a,c]	686	0.25-3.5	0.68	0.03-2	
HTF Mastigocladus[b,e]				0.3, 0.5	0.03
HTF Mastigocladus[b,f]				0.5, 1.0	<0.3
Mastigocladus laminosus[b,f]					<0.5

Synechococcus lividus[b,f]	<0.8	0.8, 1.0
Oscillatoria (6412)[a,d]	0	0.7–2.0
Oscillatoria (6506)[a,d]	0	0.67–2.0
Oscillatoria (6602)[a,d]	0	1.0–2.0
Chlorogloeopsis (6712)[a,d]	0	1.1–2.0
Pseudoanabaena (7402)[a,d]	0	1.3
Pseudoanabaena (7403)[a,d]	0	1.4–2.5
Aphanocapsa (6714)[a,d]	0	1.2–2.0
Nostoc (73102)[a,d]	0	1.5–2.7
Synechococcus (6311)[a,d]	0	0.55
Plectonema boryanum (594)[a,j]	0	0.66

[a] Garlick et al. (1977). The screening procedure included axenic cyanobacterial strains grown 5 days photoautotrophically, followed by preincubation in the presence of a Na_2S to allow for acclimatization. Experiments with nonaxenic cultures were carried out after having previously washed the cells on filters (8 μm) with the respective reaction media until no bacteria could be detected microscopically. The screening test for anoxygenic photoassimilation of CO_2 was conducted in a mineral-defined medium at the indicated sulfide concentrations, in the presence of DCMU (5 μM) and 703-nm light.

[b] Castenholtz (1976, 1977); tested field samples of almost pure cyanobacteria as judged microscopically and from pigment composition and inhibitor sensitivity. The reaction was carried out in situ in the respective hot springs after addition of the indicated sulfide concentration and DCMU (7–50 μM). Results were recorded as percent of cpm of non-DCMU nonsulfide controls and as such are qualitative.

[c] From Solar Lake (Israel).

[d] R. Y. Stainier culture collection.

[e] From hot sulfur springs, New Zealand (Castenholtz, 1976).

[f] From hot sulfur springs, U.S.A. (Castenholtz, 1977).

[g] Bardawil (Northern Sinai) salt-marsh isolate.

[h] Wadi Natrun salt-marsh isolate (Egypt).

[i] Baja (California) salt-marsh isolate.

[j] Indiana Culture Collection (Bloomington).

[k] Conducted on nonsulfide-adapted cells.

[l] Nonaxenic cultures referred to in footnote a.

and not operative at >700-nm light. Of the 30 strains examined, 16 (including *O. limnetica*) are also capable of anoxygenic photosynthesis with sulfide driven by photosystem I. Thus, photoassimilation of CO_2 has been demonstrated in the presence of sulfide, DCMU, and/or >700-nm light. Assimilation of CO_2 in the dark, in the presence or the absence of sulfide, was negligible.

Tests for photosynthetic utilization of thiosulfate by cyanobacteria are known only for *O. limnetica* (Oren, unpublished data), *S. labyrinthiformis* (Castenholtz, 1977), and *Anacystis nidulans* (Utkilen, 1976). Only in *Anacystis* has thiosulfate been shown to support phototrophic growth and to serve as an electron donor for anoxygenic photoassimilation of CO_2 in the presence of DCMU. This utilization of thiosulfate, which is oxidized to sulfate, involves adaptation of the cells and occurs only at light intensities limiting for oxygenic photosynthesis. Evolution of H_2S from either thiosulfate or sulfate has been shown in *S. lividus* (Sheridan, 1973).

Other reducing substrates such as sodium dithionite, sodium sulfite, thioglycolate, and ascorbic acid could not replace sulfide in the photoreduction of CO_2 by *S. labyrinthiformis* (Castenholtz, 1977). Molecular hydrogen, however, was shown (Belkin and Padan, 1978) to be an efficient electron donor for photoassimilation of CO_2 both in *O. limnetica* and *Aphanothece halophytica*, in an anoxygenic reaction driven by photosystem I. H_2 evolution was also shown for these species (Belkin and Padan, 1978, and unpublished data). Anoxygenic H_2 metabolism is thus shared by these nonheterocystous facultatively anoxygenic phototrophs (on sulfide) and the heterocystous types, on which cyanobacterial H_2 metabolism studies have been carried out until now (Benemann and Wear, 1974; Bothe *et al.*, 1977; Te-Or *et al.*, 1977; Weissman and Benemann, 1977).

The property of sulfide-dependent anoxygenic photosynthesis is not confined to a single cyanobacterial type (Table II) (see also Peschek, 1977). It appears in different typological groups of cyanobacteria previously defined on the basis of morphological, physiological, and biochemical properties as *Oscillatoria* and *Lyngbya* types of filamentous cyanobacteria and as an *Aphanothece* type of unicellular cyanobacteria (Stanier *et al.*, 1971; Keynon *et al.*, 1972; Rippka, 1972). The tested nonaxenic strains may possibly include additional positive types such as *Spirulina* and *Microcoleus*. No heterocystous cyanobacterium has been shown as yet to possess anoxygenic photosynthesis. Whether this is merely circumstantial, due to testing mainly nonheterocystous types, remains to be clarified. However, mainly nonheterocystous types were found in sulfide-rich habitats where *Oscillatoria* types were very prominent (Section 3; Table I) (Wood, 1965; Reynolds and Walsby, 1975; Meffert, 1975; Renaut *et al.*, 1975; Caldwell, 1977; Meffert and Krambeck, 1977; Potts and Whitton, 1977; Potts *et al.*, 1977).

The property of anoxygenic photosynthesis is not limited to particular ecosystems, geographical locations, or culture conditions (Table II). Eight strains were isolated from marine environments in different parts of the world (e.g., salt marshes in Israel and U.S.A.; Solar Lake, Israel): *O. limnetica* and *Oscillatoria salina* were collected from the sulfide-rich layers of the Solar Lake (Table II). Two positive strains, *O. amphigranulata* and *S. labyrinthiformis*, were isolated from hot springs containing sulfide. The three remaining positive strains, *Oscillatoria* 6407 and *Lyngbya* 7004 and 7104, are freshwater strains grown in culture under aerobic conditions for several years; there are no reports of their presence in natural anaerobic environments (Keynon *et al.*, 1972), but this is not surprising in view of the facultative nature of sulfide-dependent anoxygenic photosynthesis in cyanobacteria.

As in anoxygenic photosynthesis, photoheterotrophic growth of cyanobacteria in the presence of sugars and DCMU is supported by photosystem I (Rippka, 1972), but no direct correlation between these metabolic processes has been found (Garlick *et al.*, 1977).

Cultures of *A. halophytica* were studied for comparison with *O. limnetica* (Garlick *et al.*, 1977). The former also oxidizes two molecules of sulfide for each CO_2 assimilated, with elemental sulfur, the final product, being excreted from the cells. As with *O. limnetica*, an adaptation period of 1 hr in the presence of light and sulfide is needed for anoxygenic photosynthesis in *A. halophytica*.

Both the maximal rates of the anoxygenic CO_2 photoassimilation and the respective sulfide concentrations differed markedly among the strains when measured under identical incident light conditions (Table II), differences which can be attributed to variations in the cyanobacterial strains. In interpreting the results, however, the possibility of distortions due to nonoptimal experimental conditions such as limiting light intensity, temperature, or pH cannot be excluded.

A comparison of anoxygenic photosynthesis with respect to sulfide concentration could be properly carried out among *Lyngbya* 7104, *A. halophytica*, and *O. limnetica* (Garlick *et al.*, 1977) (Fig. 5b). These strains are similar in saturation intensities at 703-nm light (0.9 $\mu E/cm^2$ per min) and in their linear kinetics of anoxygenic CO_2 photoassimilation. The experimental conditions included pH 6.8, optimal temperatures, and saturating light intensities. All strains exhibited a similar general pattern of dependence on sulfide concentration—an optimum curve rather than a saturation curve. The drop in photosynthetic rates at high sulfide concentrations is due to sulfide toxicity; thus, the observed maximal rates need not be the true ones. Nevertheless, these rates are similar to that of the oxygenic photosynthesis. While the trend is similar, both the initial concentration of sulfide permitting photoassimilation and the optimum concentration for maximum rate are different in all strains. Hence, in each strain, the system utilizing the sulfide appears to differ in affinity toward sulfide, and the strains also differ in their tolerance for sulfide. Thus, each strain shows a different range of sulfide concentration at pH 6.8, at which anoxygenic photosynthesis is carried out: *Lyngbya*, 0.1–0.3 mM; *A. halophytica*, 0.1–1.2 mM; and *O. limnetica*, the highest concentration, 0.7–5 mM.

Sulfide is a weak acid with pK values of 6.8 and 11.1 (in the ionic strength of the *O. limnetica* growth medium), and the proportion of ionized and non-ionized sulfide species will vary with the pH of the medium. A marked inhibition of anoxygenic photosynthesis in the presence of 4 mM sulfide was observed (Oren and Padan, unpublished data) when reducing the external pH from pH 7 to pH 6 (Fig. 5c). Oxygenic photosynthesis was not affected by a similar change in pH. Comparable pH effects were observed with different sulfide concentrations (2–12 mM); the higher the sulfide concentration, the higher the pH at which the inhibition was expressed. This pH effect could be related to the nonionized intracellular H_2S, without regard to the total medium sulfide concentrations. Intracellular H_2S was calculated by assuming that the cell membrane is permeable only to this nonionized sulfide species and that the cytoplasmic pH is a constant 7.6 at this external pH range, as in other cyanobacteria (Padan and Schuldiner, 1978) and bacteria (Padan *et al.*, 1976).

Tests for anaerobic growth which are positive in *O. limnetica* have yielded negative results in *A. halophytica*. Accordingly, *A. halophytica* contains polyunsaturated fatty acids and an O_2-requiring desaturation mechanism (Oren, Devir, and Padan, 1978, unpublished data), whereas *O. limnetica* possesses anaerobic fatty acid metabolism (see Section 4.4). Thus, anoxygenic photosynthesis need not necessarily be linked with anaerobic growth and anaerobic fatty acid metabolism. It is suggested that these physiological characteristics, possibly representing different degrees of anaerobism, may serve as taxonomic markers for extension of the recently commenced reclassification of cyanobacteria (Stanier *et al.*, 1971; Kenyon, 1972; Kenyon *et al.*, 1972). For example, *Spirulina* sp., possibly related to the facultatively anoxygenic species, *S. labyrinthiformis*, also possesses the anerobic fatty acid metabolism of *O. limnetica* (Kenyon *et al.*, 1972).

Sulfide appears more toxic to oxygenic photosynthesis in the cyanobacteria than to anoxygenic photosynthesis (Table II). In *O. limnetica*, the former is inhibited by 0.1 mM sulfide, whereas the latter is inhibited only by >4 mM (at pH 6.8). These differences are also observed in other facultatively anoxygenic strains (Table II). An exception to this may be *O. terebriformis*, which seems to tolerate a high (0.8 mM) sulfide concentration in oxygenic photosynthesis. However, no details were given on the pH, purity of the population, etc., in this case (Castenholtz, 1977). The concentration of sulfide which is toxic for cyanobacteria that are not capable of anoxygenic photosynthesis is also within the low concentration range (<0.5 mM) (Table II).

The mechanism of sulfide toxicity is not at all clear in either phototrophic or heterotrophic metabolism. In different organisms, phototrophs as well as heterotrophs, the effects of sulfide are paralleled to a great extent by those of cyanide (Evans, 1967; Izawa and Good, 1972). Both act as inhibitors of catalases, peroxidases, succinic dehydrogenase, carbonic anhydrase, cytochrome oxidase, and other enzymes, and both sulfide and cyanide tend to form complexes with heme iron. Cyanide is a potent inhibitor of CO_2 fixation in many phototrophs, probably through the poisoning of ribulose diphosphate carboxylase and the primary carboxylation reaction (Good and Izawa, 1973).

In cyanobacteria, the apparently greater sensitivity to sulfide in oxygenic photosynthesis as compared with anoxygenic photosynthesis suggests that photosystem II is more susceptible to sulfide poisoning than photosystem I. In a recent study of the effect of sulfide on the variable fluorescence of photosystem II in *O. limnetica*, Oren, Padan, and Malkin (1978, unpublished data) showed that sulfide inhibits electron transport from water to photosystem II. The anoxygenic photosynthetic property operating with photosystem I allows for greater sulfide tolerance and thus confers a selective advantage over the nonanoxygenic photosynthetic cyanobacteria in sulfide-rich habitats. This is true even for facultatively anoxygenic photosynthesizing cyanobacteria strains which cannot grow under purely anaerobic conditions. These strains will maintain themselves under sulfide-rich conditions and, by oxidizing sulfide, will decrease the sulfide concentration in their surroundings. However, the most successful strains will be those with facultatively anaerobic growth capable of anoxygenic photosynthesis, as in *O. limnetica*.

The differences in sulfide tolerance encountered with facultatively anoxygenic cyanobacteria are known in photosynthetic sulfur bacteria and often constitute a determinative factor in their ecology (Baas Becking and Wood, 1955; Van Niel, 1963; Pfennig, 1975). Thus, the Chlorobiaceae photosynthesize at the highest sulfide concentrations of 4–8 mM, as *O. limnetica*, whereas Chromatiaceae do so at 0.8–4 mM, and Rhodospirillaceae at 0.4–2 mM (Pfennig, 1975). In addition, the extracellular sulfur deposition of *O. limnetica* and *A. halophytica* is similar to that of the sulfide-oxidizing photosynthetic bacteria, the Chlorobiaceae, and some Chromatiaceae (Pfennig, 1977). However, the pattern of sulfide oxidation of the cyanobacteria strains seems thermodynamically inefficient; while photosynthetic bacteria are capable of oxidizing sulfide to sulfate (Pfennig and Trüper, 1974; Pfennig, 1975, 1977), and gain 8 electrons for each H_2S molecule, *O. limnetica* and *A. halophytica* can only remove 2 electrons per molecule. The oxidation of sulfide to sulfur seems to be a by-product in the sulfide-oxidation process in photosynthetic sulfur bacteria (Trüper, 1973). Furthermore, *O. limnetica*, whether sulfide induced or oxygenic photoautotrophic, and *S. labyrinthiformis* (Castenholtz, 1977) could not be shown to use thiosulfate as an electron donor for CO_2 photoassimilation (see

Section 5.1), while this substrate and other sulfur-containing compounds are used efficiently by many sulfide-oxidizing photosynthetic bacteria (Roy and Trudinger, 1970; Pfennig and Trüper, 1974; Pfennig, 1975). As mentioned above, the ionized HS^- and S^{2-}, and the nonionized H_2S species, will be present in reaction mixtures in different proportions according to the pH and the respective pK values. In addition, the sulfide species which is taken up, serves as a substrate, or is toxic to the cell need not be the same. Hence, a marked effect of external pH on anoxygenic photosynthesis, shown for *O. limnetica* (Fig. 5c) and also in photosynthetic bacteria (Baas Becking and Wood, 1955), is expected, though the pH effect need not be identical in all sulfide-utilizing species. All these observed and expected variabilities among the anoxygenic photosynthesizing phototrophs should be expressed at the ecological level, leading to different distribution patterns in the sulfide-concentration gradients prevailing in the anaerobic habitats. This is indeed realized in the Solar Lake (Israel) and sulfur-containing hot springs (New Zealand and U.S.A.) (see Sections 5.2 and 5.3), where the expression of facultatively anoxygenic photosynthesis of cyanobacteria at the ecological level has been thoroughly investigated (Castenholtz, 1976, 1977; Cohen *et al.*, 1977a,b).

5.2. Facultatively Anoxygenic Photosynthesis of Cyanobacteria in Hot Sulfur Springs

Castenholtz (1976, 1977) (Fig. 1; Tables I and II; Sections 2.1 and 3.1) investigated phototrophic capacities of cyanobacteria forming the continuum along the sulfide gradient of hot springs in New Zealand and Yellowstone (U.S.A.), described above. Tests carried out *in situ* with field populations can explain their ecological distribution. Only *O. amphigranulata* and *S. labyrinthiformis*, which do occur at the highest sulfide concentrations, can carry out anoxygenic photosynthesis *in situ* in the presence of high sulfide concentrations. Strains occurring further downstream are inhibited by much lower sulfide concentrations. Additions of sulfide to concentrations of >0.6 mM resulted in an increased photoincorporation of CO_2 in *O. amphigranulata* growing in New Zealand hot springs; these photosynthesis rates were commonly more than three times those of the nonsulfide controls, and the rates with 50 μM DCMU and sulfide were usually twice those of the nonsulfide, non-DCMU controls (Castenholtz, 1976). Furthermore, without sulfide, the rates were very low when compared to the same or to different species growing in low sulfide concentrations or in nonsulfide waters. Hence, a great extent of the phototropic activity of this cyanobacterium in its natural habitat appears to be anoxygenic with sulfide serving as the electron donor.

Natural *O. amphigranulata* from springs with high sulfide concentrations are colored dull green to dull yellow-green (Castenholtz, 1976) due to low amounts of the photosystem II light-harvesting pigments, C-phycocyanin and C-phycoerythrin. The pigment composition accounts for the low rates of oxygenic photosynthesis of these populations and suggests a differentiation toward anoxygenic photosynthesis. The pigment composition of *O. amphigranulata* appeared normal in water with very low sulfide content or without sulfide; these cells are incapable of photoassimilating CO_2 in the presence of very low sulfide concentrations. Thus, acclimation to sulfide appears to take place in *O. amphigranulata* in natural habitats, as in *O. limnetica* in culture. No differences in pigment composition were found between oxygenic and anoxygenic cells in *O. limnetica* (Oren et al., 1977; Oren and Padan, 1978) (Section 4.3); however, adaptation may possibly involve changes in pigment composition in *O. amphigranulata*. This field observation needs confirmation from investigations with pure cultures under defined experimental conditions, as differences observed between the sufide-rich and nonsulfide field populations may be due to unrelated causes. In any event, the sufide tolerance and the capacity of *O. amphigranulata* for anoxygenic photosynthesis confers an ecological advantage in its major habitat. For example, *HTF Mastigocladus*, which shares the downstream regions with *O. amphigranulata* and other cyanobacteria, is completely excluded from high-sulfide areas, and even low concentrations of sulfide (0.03 mM) inhibit photoassimilation of CO_2. The experiment was done with *HTF Mastigocladus* from nonsulfide spring water or from water containing a sulfide concentration tolerated by this species.

A similar observation was made on populations of *M. laminosus, HTF M. laminosus*, and *S. labyrinthiformis* from Yellowstone Park springs occurring in nonsulfide and high-sulfide waters (Castenholtz, 1977). Only the last species appears to adapt to high sulfide concentrations in its natural environment. Photoassimilation of CO_2 *in situ* in *S. labyrinthiformis* strains from sulfide-rich upstream waters was slightly stimulated by sulfide (0.6–1 mM) and continued in the presence of 7 µM DCMU or >700-nm light. Without sulfide, DCMU or these light wavelengths completely inhibited the oxygenic photosynthesis. Sulfide also relieved the inhibitory effects of hydroxylamine, an inhibitor of the O_2-evolving system. *Spirulina* also appears to shift readily to oxygenic photosynthesis when sulfide is removed (Castenholtz, 1977), as *O. limnetica* is known to do (Oren and Padan, 1978).

5.3. Facultatively Anoxygenic Photosynthesis of Cyanobacteria in a Stratified Lake

The Solar Lake, a small (140 × 50 m), shallow sea-margin pond with a depth of 4–6 m, depending on the time of the year, is located 18 km south of Elat on the Sinai coast of the Gulf of Elat (Red Sea); it is separated from the Gulf by a 60-m-wide gravel bar. The unique limnological features of this small water body (Cohen *et al.*, 1977a,b) are determined by water seepage from the Gulf to the pond surface, the arid climatic conditions, and wind protection afforded by the surrounding mountain ridge. It is a mesothermic, hypersaline desert lake with extremely high heliothermal heat accumulation and with an unusual type of monomixis, a summer overturn.

During fall, halocline and pycnocline are established by seepage of sea water, which is lower in salinity than the Solar Lake water, into the lake surface layer. As a result of this density stratification, heliothermal heating of the lower water layer results in an inverse thermal profile with a thermocline at 18°C/m with extreme differences in conditions. The stratified bottom layers are very abundant in sulfide, which reaches a concentration of 1.5 mM (Fig. 2). Stratification breaks down in early summer when holomixis occurs due to the very high insolation. At this time, the inflow of seawater no longer compensates for the high evaporation rate. Holomixis lasts for several weeks, when the water reaches a temperature of ca. 29°C, a pH of 8.5–8.7 (Cohen *et al.*, 1977a), and an O_2 content of 4–5 mg/liter (Cohen *et al.*, 1977b).

During the winter stratification, the photosynthetic community (Fig. 8) in the upper, oxygenated layer is composed mainly of low concentrations of diatoms and a unicellular cyanobacterium, *A. halophytica*. At the O_2-sulfide borderline, there is a plate formed by the purple photosynthetic bacterium

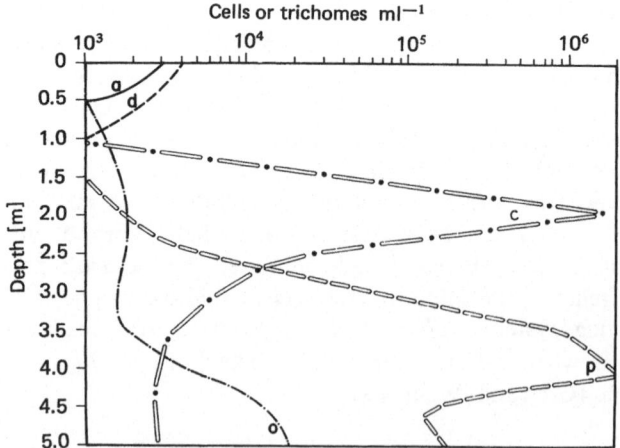

Figure 8. Vertical distribution of photosynthetic microorganisms during stratification. Numbers (direct counts) of each group from samples taken on March 27, 1971. d, Diatoms (*Nitzschia* sp., *Amphora* sp., *Navicula* sp.); a, unicellular cyanobacteria (*Aphanothece halophytica*, *Aphanocapsa littoralis*); c, *Chromatium violescens*; p, *Prosthecochloris* sp.; o, filamentous cyanobacteria (*Oscillatoria limnetica*, *Oscillatoria salina*, *Microcoleus* sp.). From Cohen *et al.* (1977b).

Chromatium violescens; below it, at still higher concentrations of sulfide, there is a layer with the green bacterium *Prosthecochloris* sp. In the deepest layers with the highest sulfide concentrations, *O. limnetica* proliferates; other cyanobacteria species, *Microcoleus* sp., and *O. salina* are also present. The bottom layers reach 30 g C/m^3. Primary productivity measurements (Cohen *et al.*, 1977b) in these layers yielded very high rates (4960 mg C/m^3 per day), higher than the values encountered for other nonpolluted water bodies. This photosynthetic activity appears to be anoxygenic, as it is not inhibited by DCMU (Garlick, unpublished data); also, the light wavelengths penetrating into the lake which are not absorbed by photosynthesizing bacteria, between 500 and 600 nm, are more effective in driving anoxygenic photosynthesis than oxygenic photosynthesis (Section 4.3) (Oren *et al.*, 1977). From the rate of primary productivity and the standing crop, a doubling time of 1–2 days was calculated, a very similar growth rate to that measured for *O. limnetica* under anaerobic culture conditions (Section 4.4) (Oren and Padan, 1978).

The stratification pattern of phototrophs in the anoxygenic layers of the Solar Lake corresponds with their optimum sulfide concentration range (van Niel, 1963; Pfennig, 1975). Purple photosynthetic bacteria, tolerant only of low sulfide concentrations, thrive at the O_2 borderline in the Solar Lake. Green bacteria, with a much higher optimum sulfide range, flourish in the bottom layers along with *O. limnetica*, which photosynthesizes anoxygenically in 1–4 mM sulfide (at pH 6.8).

O. limnetica peaks, however, below the *Prostecochloris* level and thus appears incapable of competing with the green bacteria. This could be due to the *O. limnetica* requirement for a higher sulfide concentration for optimal anoxygenic photosynthesis, whereas *Prostecochloris* may require less. In addition, it should be remembered that the process of sulfide oxidation by *O. limnetica*, with sulfide oxidized only to elemental sulfur (Section 4.3) (Cohen *et al.*, 1975b), is thermodynamically inefficient in comparison to that of the photosynthetic bacteria. This inefficiency may also contribute to the inferiority of *O. limnetica* in the presence of green bacteria.

The ecological distribution of phototrophic anaerobes in the Solar Lake, including all types of sulfide-utilizing phototrophs along the density stratification, allows efficient utilization of the available light energy. The purple bacteria, utilizing the longest wavelengths (700–800 nm), occur above the green bacteria, utilizing shorter wavelengths (600–700 nm). The cyanobacteria, which utilize the deep-penetrating short wavelengths in anoxygenic photosynthesis (500–600 nm), occur at the deepest layers. Furthermore, the primary production of the stratified Solar Lake is almost entirely restricted to the metalimnion and hypolimnion. Since stratification lasts for at least 9 months, the contribution of anoxygenic photosynthesis to the annual production is as high as 82.3% (Cohen *et al.*, 1977b).

A. halophytica, with an optimum for anoxygenic photosynthesis at 0.7 mM sulfide (pH 6.8) (Fig. 5b), is absent from the bottom layers, but it is also absent from the sulfide–O_2 hypolimnion borderline with an optimum sulfide concentration for its anoxygenic photosynthesis (Section 5.1), a layer in which *Chromatium* thrives. In addition to the sulfide gradient, the hypolimnion of the Solar Lake is characterized by extremely high temperatures (50–60°C), a range prohibitive for *A. halophytica* growth. Since *A. halophytica* is capable of oxygenic photosynthesis, it occurs in the upper oxygenated layers. However, its capacity for anoxygenic photosynthesis permits survival during temporary, brief exposure to the neighboring anaerobic layers.

During the spring-to-summer overturn period, the water column becomes aerobic, the phototrophic bacteria disappear, and *O. limnetica* shifts to its oxygenic metabolism and thrives in the water column along with *A. halophytica*. At that time, eucaryotic algae are also present throughout the lake. As they are only capable of oxygenic photosynthesis, they disappear at the next anaerobic period, while *O. limnetica* shifts to anaerobic phototrophic metabolism and continues to flourish. Hence, *O. limnetica*, by utilizing combined oxygenic and anoxygenic photosynthetic capacities, is the dominant phototroph species of the Solar Lake, where photoanaerobic and photoaerobic conditions alternate with time.

6. Concluding Remarks; Ecological and Evolutionary Importance of Photoanaerobic Metabolism in Cyanobacteria

The property of facultatively anoxygenic photosynthesis, which is widespread among cyanobacteria, is a significant factor in the ecology of cyanobacteria of the Solar Lake and of alkaline and neutral hot sulfur springs. In these ecosystems, in which sulfide-rich photoanaerobic and photoaerobic conditions alternate with time as well as with distance downstream, cyanobacteria with facultatively anoxygenic phototrophic physiology predominate. The facultatively anoxygenic phototrophic capability may be shared by numerous other cyanobacteria species found in different habitats with alternating photoaerobic and photoanaerobic conditions, e.g., cyanobacteria of the metalimnion of stratified lakes, of poly-mictic "cyanobacteria" equatorial lakes, and of the marine "thiobioses," etc. (see Sections 2 and 3). Therefore, it is highly suggestive that the significance of facultatively anoxygenic phototrophy is of paramount importance in the global ecology of the cyanobacteria.

Cyanobacteria have long been known to be the only photosynthetic N_2-fixing organisms under aerobic conditions. These include mainly heterocystous types (Fogg *et al.*, 1973), with the exception of *Gloeocapsa* (Gallon *et al.*, 1975) and *Trichodesmium* (Carpenter and Price, 1977). Aerobic fixation most probably takes place in the heterocysts (Fay *et al.*, 1968), though under microaerophilic conditions; N_2 fixation is often even more efficient (Stewart and Pearson, 1970). Recently, vegetative cells of heterocystous cyanobacteria and also of nonheterocystous ones have also been shown to be capable of anaerobic (or microaerophilic) N_2 fixation (Stewart and Lex, 1970; Rippka and Waterbury, 1977). Given the apparent ubiquity of cyanobacteria in photoanaerobic habitats, it may be presumed that their contribution to the N_2 balance in these ecosystems is significant. Thus, anaerobic fixation by cyanobacteria, in addition to the aerobic N_2 fixation, may have a great economic impact in ecosystems such as rice paddies (Fogg *et al.*, 1973; Stewart, 1973; Renaut *et al.*, 1975; Venkataraman, 1975). N_2 fixation in nonheterocystous cyanobacteria has recently been documented in anaerobic sediments (Kjeldsen and Walsby, 1977; Potts and Whitton, 1977; Potts *et al.*, 1977). Additional surveys are urgently needed to quantify anaerobic N_2 fixation by cyanobacteria in nature, with particular reference to the facultatively anoxygenic nonheterocystous types.

Some of the phototrophic physiological characteristics of the facultatively anoxygenic photosynthetic cyanobacteria are encountered in cyanobacteria that do not carry out anoxygenic photosynthesis. Many cyanobacteria tolerant of low sulfide concentrations exhibit actual preference for these concentrations. Thus, Stewart and Pearson (1970) showed that *Anabaena flos-aquae* is incapable of anoxygenic photosynthesis, i.e., in the presence of DCMU and sulfide, while its photoassimilation rate is improved in the presence of sulfide when photolysis of water (in the presence of salicylaldoxime) is impaired but not completely inhibited. As sulfide readily reacts with O_2, this sulfide effect may be due to a reduction in deleterious effects of O_2 noted in cyanobacteria (Stewart and Pearson, 1970; Aveliovich and Shilo, 1972), as well as to beneficial reducing conditions. In addition, it has been suggested that under these specific conditions, sulfide may also partially serve as an electron donor. The beneficial sulfide effect on CO_2 photoassimilation has also been shown by Weller *et al.* (1975) with a species of *Phormidium*; however, this increase in photoassimilation is due to a reduction in the redox potential of the system. Hence, reducing agents other than sulfide, such as cysteine, thioglycolate, and sulfite, had a stimulatory effect above the level seen in an N_2 atmosphere.

Preference for low O_2 tension and/or redox potential appears to be a general characteristic of many cyanobacteria, and Fogg *et al.* (1973) concluded, "it is likely that in the light blue-green algae in general grow more rapidly under microaerophilic than under fully aerobic conditions" (see also Allen, 1952; Gusev, 1962; Topachevskii *et al.*, 1969; Stewart and Pearson, 1970; Stewart, 1973). Possession of this property accounts for the selective advantage of cyanobacteria in an ecosystem where alternations between aerobic and anaerobic conditions do not involve drastic sulfide changes, as in eutrophic epilimnion. It is also possible that this property of cyanobacteria underlies the growth pattern of cyanobacteria in nature, whether as symbionts or as free-living forms in bundles, colonies, nodes, or in dense blooms (Fogg *et al.*, 1973; Weller *et al.*, 1975; Caldwell, 1977). All these forms of life may represent adaptation to microaerophilic microenvironments created either by shading, by rich bacterial growth, and/or by accumulation of reducing substances (Caldwell, 1977).

Oxygenic photosynthesis, on the one hand, and facultatively anoxygenic phototrophic capacity along with preference for low redox potential and microaerophilic conditions, on the other hand, permit cyanobacteria to occupy a more intermediate ecological and physiological position than that made possible by the strictly anaerobic phototrophic physiology of photosynthetic bacteria or the strictly aerobic physiology of eucaryotic phototrophs. This interlinking position of cyanobacteria in the phototrophic world is compatible with the fact that the cyanobacteria are procaryotes and among the oldest organisms on earth, occurring in Precambrian rock and perhaps even earlier (Schopf, 1974); this leads to speculation on the evolutionary pattern of the phototrophs.

If evolution of chlorophyll a preceded bacterial chlorophyll, as suggested by Olson (1978), and if photosynthetic organisms had to evolve in the presence of sulfide and limiting light intensities under large columns of water in order to avoid destructive irradiation, primary competition for the spectrum should have yielded organisms such as photosynthetic bacteria occupying layers above anoxygenic cyanobacteria. Anoxygenic cyanobacteria which retained the chlorophyll a and, in addition, also developed phycobiliproteins could more efficiently utilize the more deeply penetrating short-wavelength light and therefore could have remained in greater depths. Such a pattern of light-spectrum utilization by phototrophs is observed during winter in the sulfide bottom layer of the Solar Lake, the ecotope of *O. limnetica* (Section 5.3). In this respect, it is also noteworthy that two of the sulfide-rich ecosystems abundant in cyanobacteria, the marine "thiobiosis" and the sulfur hot springs, may represent old ecosystems, perhaps preceding the oxidized biosphere in age. By retaining chlorophyll a, the cyanobacteria retained the anoxygenic option with photosystem I while also developing photosystem II, the origin of plant-type oxygenic photosynthesis. Moreover, it is highly possible that cyanobacteria represent the predecessors of the contemporary chloroplasts of eucaryotic plant cells (Margulis, 1968; Stanier, 1974). Whether this scheme of evolution of phototrophy is correct or not is clearly open to question. Yet the discovery of the unique cyanobacterial facultatively anoxygenic photoautotrophic physiology clarifies certain hitherto completely obscure aspects of cyanobacterial ecology, the occurrence and predominance of cyanobacteria in habitats alternating between photoaerobic and photoanaerobic conditions.

ACKNOWLEDGMENTS

The author thanks Prof. M. Shilo and A. Oren, Department of Microbiological Chemistry, Hebrew University, Jerusalem, Israel, and Prof. W. E. Krumbein,

University of Oldenburg, Oldenberg, West Germany, for critical reading of the manuscript, and Dr. N. Ben-Eliahu for editorial assistance. Writing of this review was supported by the German Federal Ministry for Research and Technology (GKSS).

References

Abeliovich, A., 1967, Water blooms of blue green algae and oxygen regime in fish ponds, *Verh. Int. Ver. Limnol.* 17:594–601.

Abeliovich, A., and Shilo, M., 1972, Photooxidative death in blue-green algae, *J. Bacteriol.* 111:682–689.

Allen, M. B., 1952, The cultivation of Myxophyceae, *Arch. Mikrobiol.* 17:34–53.

Anagnostidis, K., and Golubić, S., 1966, Über die Ökologie einiger *Spirulina*-Arten, *Nova Hedwigia* 11:309–335.

Andreasen, A. A., and Stier, T. J. B., 1953, Anaerobic nutrition of *Saccharomyces cerevisiae*, *J. Cell. Comp. Physiol.* 41:23–36.

Arnon, D. I., Losada, M., Nozaki, M., and Tagawa, K., 1961, Photoproduction of hydrogen, photofixation of nitrogen and a unified concept of photosynthesis, *Nature (London)* 190:601–606.

Avron, M., 1977, Energy transduction in chloroplasts, *Annu. Rev. Biochem.* 46:143–155.

Baas Becking, L. G. M., and Wood, E. J. F., 1955, Biological processes in the estuarine environment. I–II. Ecology of the sulphur cycle, *Proc. Acad. Sci. Amst. B* 58:160–181.

Baxter, R. M., Prosser, M. V., Talling, J. F., and Wood, R. B., 1965, Stratification in tropical African lakes at moderate altitudes (1500 to 2000 m), *Limnol. Oceanogr.* 10:510–520.

Belkin, S., and Padan, E., 1978, Hydrogen metabolism in the facultative anoxygenic cyanobacteria (blue-green algae) *Oscillatoria limnetica* and *Aphanothece halophytica*, *Arch. Microbiol.* 116:109–111.

Benemann, J. R., and Wear, N. M., 1974, Hydrogen evolution by nitrogen fixing *Anabaena cylindrica* cultures, *Science* 184:174–175.

Berman, T., and Pollingher, U., 1974, Annual and seasonal variations of phytoplankton, chlorophyll, and photosynthesis in Lake Kinneret, *Limnol. Oceanogr.* 19:31–54.

Biebl, H., and Pfennig, N., 1977, CO_2-fixation by anaerotic phototrophic bacteria in lakes, in: *Workshop on the Measurement of Microbial Activities in the Carbon Cycle in Freshwaters and Standardization of Methods*, Max Planck Institute of Limnology, Plön (August 4–5, 1977).

Blackburn, T. H., Kleiber, P., and Fenchel, T., 1975, Photosynthetic sulfide oxidation in marine sediments, *Oikos* 26:103–108.

Bloch, K., Baronowski, Y., Goldfine, H., Lennarz, W. J., Light, R., Norris, A. T., and Scheuerbrandt, G., 1966, Biosynthesis and metabolism of unsaturated fatty acids, *Fed. Proc. Fed. Am. Soc. Exp. Biol.* 20:921–927.

Bothe, H., Tennigkeit, J., Eisenberger, G., and Yates, M. G., 1977, The hydrogenase–nitrogenase relationship in the blue-green alga *Anabaena cylindrica*, *Planta* 133:237–242.

Bozniak, E. G., and Kennedy, L. L., 1968, Periodicity and ecology of the phytoplankton in an oligotrophic and eutrophic lake, *Can. J. Bot.* 46:1259–1271.

Brock, T. D., 1973a, Evolutionary and ecological aspects of the cyanophytes, in: *The Biology of Blue-Green Algae* (N. G. Carr and B. A. Whitton, eds.), pp. 487–500, Blackwell Scientific Publications, Oxford.

Brock, T. D., 1973b, Lower pH limit for the existence of blue-green algae: Evolutionary and ecological implications, *Science* 179:480–483.

Brook, A. J., Baker, A. L., and Klemer, A. R., 1971, The use of turbidometry in studies of the population dynamics of phytoplankton populations, with special reference to *Oscillatoria agardhii* var. *isothrix, Mitt. Int. Ver. Theor. Angew. Limnol.* 19:244–252.

Burgis, M. J., Darlington, J. P. E. C., Dunn, I. G., Ganf, G. G., Gwahaba, J. J., and McGowan, L. M., 1973, The biomass and distribution of organisms in Lake George, Uganda, *Proc. R. Soc. London Ser. B Biol. Sci.* 184:271–298.

Caldwell, D. E., 1977, The planktonic microflora of lakes, *CRC Crit. Rev. Microbiol.* 5:305–370.

Carpenter, E. J., 1973, Nitrogen fixation by *Oscillatoria (Trichodesmium) thiebautii* in the southwestern Sargasso Sea, *Deep-Sea Res.* 20:285–288.

Carpenter, E. J., Price, C. C., 1977, Nitrogen fixation, distribution, and production of *Oscillatoria (Trichodesmium)* spp. in the western Sargasso and Caribbean Seas, *Limnol. Oceanogr.* 22:60–72.

Castenholtz, R. W., 1969, Thermophilic blue-green algae and the thermal environment, *Bacteriol. Rev.* 33:476–504.

Castenholtz, R. W., 1973, The possible photosynthetic use of sulfide by the filamentous phototrophic bacteria of hot springs, *Limnol. Oceanogr.* 18:863–876.

Castenholtz, R. W., 1976, The effect of sulfide on the blue-green algae of hot springs. I. New Zealand and Iceland, *J. Phycol.* 12:54–68.

Castenholtz, R. W., 1977, The effect of sulfide on the blue-green algae of hot springs. II. Yellowstone National Park, *Microb. Ecol.* 3:79–105.

Cohen, Y., Padan, E., and Shilo, M., 1975a, Facultative anoxygenic photosynthesis in the cyanobacterium *Oscillatoria limnetica, J. Bacteriol.* 123:855–861.

Cohen, Y., Jørgensen, B. B., Padan, E., and Shilo, M., 1975b, Sulfide-dependent anoxygenic photosynthesis in the cyanobacterium *Oscillatoria limnetica, Nature (London)* 257:489–492.

Cohen, Y., Krumbein, W. E., Goldberg, M., and Shilo, M., 1977a, Solar Lake (Sinai). 1. Physical and chemical limnology, *Limnol. Oceanogr.* 22:597–608.

Cohen, Y., Krumbein, W. E., and Shilo, M., 1977b, Solar Lake (Sinai). 2. Distribution of photosynthetic microorganisms and primary production, *Limnol. Oceanogr.* 22:609–620.

Cohen, Y., Krumbein, W. E., and Shilo, M., 1977c, Solar Lake (Sinai). 4. Stromatolitic cyanobacterial mats, *Limnol. Oceanogr.* 22:635–656.

Cohen-Bazire, G., Sistrom, W. R., and Stanier, R. Y., 1957, Kinetic studies of pigment synthesis by non-sulfur purple bacteria, *J. Cell. Comp. Physiol.* 49:25–68.

Cramer, W. A., and Whitmarsh, J., 1977, Photosynthetic cytochromes, *Annu. Rev. Plant Physiol.* 28:133–142.

Culver, D. A., and Brunskill, G. J., 1969, Fayetteville Green Lake, New York. V. Studies of primary production and zooplankton in a meromictic marl lake, *Limnol. Oceanogr.* 14:862–873.

Czurda, V., 1940/1941, Schwefelwasserstoff als ökologischer Faktor der Algen, *Zentralbl. Bakteriol. Parasitenk. Infektionskr. Hyg. Abt. 2* 103:285–311.

Damas, H., 1964, Le plancton de quelques lacs d'Afrique centrale, *Verh. Int. Ver. Limnol.* 15:128–138.

Doemel, W. N., and Brock, T. D., 1970, The upper temperature limit of *Cyanidium caldarium, Arch. Mikrobiol.* 72:326–332.

Doemel, W. N., and Brock, T. D., 1974, Bacterial stromatolites: Origin of laminations, *Science* 184:1083–1085.

Dugdale, R. C., Menzel, D. W., and Ryther, J. H., 1961, Nitrogen fixation in the Sargasso Sea, *Deep-Sea Res.* 7:298–300.

Duthie, H. C., and Carter, J. C. H., 1970, Meromixis in Sunfish Lake, Southern Ontario, *J. Fish. Res. Board of Can.* 27:847–856.

Eberly, W. R., 1964, Primary production in the metalimnion of McLish Lake (Northern Indiana), an extreme plus-heterograde lake, *Verh. Int. Ver. Theor. Angew. Limnol.* 15:394–401.

Elster, H. J., 1966, Absolute and relative assimilation rates in relation to phytoplankton populations, in: *Primary Productivity in Aquatic Environments* (C. R. Goldman, ed.), pp. 77–103, Memorie dell'Istituto Italiano di Idrobiologia, 18th Suppl., University of California Press, Berkeley.

Evans, C. L., 1967, The toxicity of hydrogen sulfide and other sulfides, *Q. J. Exp. Physiol.* 52:231–248.

Fay, P., Stewart, W. D. P., Walsby, A. E., and Fogg, G. E., 1968, Is the heterocyst the site of nitrogen fixation in blue-green algae? *Nature (London)* 220:810–812.

Fenchel, T., 1971, The reduction–oxidation properties of marine sediments and the vertical distribution of the microfauna, in: *Vie et Milieu, Troisième Symposium Européen de Biologie Marine*, Suppl. No. 22, pp. 509–521.

Fenchel, T. M., and Riedl, R. J., 1970, The sulfide system: A new biotic community underneath the oxidized layer of marine sand bottoms, *Mar. Biol.* 7:255–268.

Fenchel, T., and Straarup, B. J., 1971, Vertical distribution of photosynthetic pigments and the penetration of light in marine sediments, *Oikos* 22:172–182.

Findenegg, I., 1966, Factors controlling primary productivity especially with regard to water replenishment, stratification and mixing, in: *Primary Productivity in Aquatic Environments* (C. R. Goldman, ed.), pp. 105–119, Memorie dell'Istituto Italiano di Idrobiologia, 18th Suppl., University of California Press, Berkeley.

Findenegg, I., 1971, Die Produktionsleistungen einiger planktischer Algenarten in ihrem natürlichen Milieu, *Arch. Hydrobiol.* 69:273–293.

Fischer, U., and Trüper, H. G., 1977, Cytochrome c-550 of *Thiocapsa roseopersicina*: Properties and reduction by sulfide, *FEMS Microbiol. Lett.* 1:87–90.

Fogg, G. E., and Walsby, A. E., 1971, Buoyancy regulation and the growth of planktonic blue-green algae, *Mitt. Int. Ver. Theor. Angew. Limnol.* 19:182–188.

Fogg, G. E., Stewart, W. D. P., Fay, P., and Walsby, A. E., 1973, *The Blue-Green Algae*, Academic Press, London and New York.

Frenkel, A. W., 1970, Multiplicity of electron transport reactions in bacterial photosynthesis, *Biol. Rev.* 45:569–616.

Fritsch, F. E., 1907, A general consideration of the subaerial and fresh-water algal flora of Ceylon. A contribution to the study of tropical algal ecology. Part I. Subaerial algae and algae of the inland fresh-waters. *Proc. R. Soc. London Ser. B Biol. Sci.* 179:197–254.

Fritsch, F. E., 1945, *The Structure and Reproduction of Algae*, Vol. II, Cambridge University Press, Cambridge.

Gaffron, H., 1944, Photosynthesis, photoreduction and dark reduction of carbon dioxide in certain algae, *Biol. Rev.* 19:1–20.

Gallon, J. R., Kurz, W. G. W., and LaRue, T. A., 1975, The physiology of nitrogen fixation by a *Gloeocapsa* sp., in: *Nitrogen Fixation by Free-Living Micro-organisms* (W. D. P. Stewart, ed.), pp. 159–173, Cambridge University Press, Cambridge.

Ganf, G. G., 1974a, Incident solar irradiance and underwater light penetration as factors controlling the chlorophyll a content of a shallow equatorial lake (Lake George, Uganda), *J. Ecol.* 62:593–610.

Ganf, G. G., 1974b, Diurnal mixing and the vertical distribution of phytoplankton in a shallow equatorial lake (Lake George, Uganda), *J. Ecol.* 62:611–629.

Ganf, G. G., 1974c, Phytoplankton biomass and distribution in a shallow eutrophic lake (Lake George, Uganda), *Oecologia* 16:9–29.

Ganf, G. G., and Horne, A. J., 1975, Diurnal stratification, photosynthesis and nitrogen-fixation in a shallow, equatorial lake (Lake George, Uganda), *Freshwater Biol.* 5:13–39.

Ganf, G. G., and Viner, A. B., 1973, Ecological stability in a shallow equatorial lake (Lake George, Uganda), *Proc. R. Soc. London Ser. B Biol. Sci.* 184:321–346.

Garlick, S., Oren, A., and Padan, E., 1977, Occurrence of facultative anoxygenic photosynthesis among filamentous and unicellular cyanobacteria, *J. Bacteriol.* 129:623–629.

Gest, H., 1972, Energy conversion and generation of reducing power in bacterial photosynthesis, *Adv. Microb. Physiol.* 7:243–282.

Gibbs, M., 1962, Fermentation, in: *Physiology and Biochemistry of Algae* (R. A. Lewin, ed.), pp. 91–97, Academic Press, New York and London.

Goad, L. J., and Goodwin, T. W., 1972, Biosynthesis of plant sterols, in: *Progress in Phytochemistry* (L. Reinhold and Y. Liwschitz, eds.), Vol. 3, pp. 113–198, John Wiley–Interscience, London.

Golubić, S., 1973, The relationship between the blue-green algae and carbonate deposits, in: *The Biology of Blue-Green Algae* (N. G. Carr and B. A. Whitton, eds.), pp. 434–472, Blackwell Scientific Publications, Oxford.

Good, N. E., and Izawa, S., 1973, Inhibition of photosynthesis, in: *Metabolic Inhibitors* (R. N. Hochster, M. Kates, and J. H. Quastel, eds.), pp. 179–214, Academic Press, New York.

Gorham, E., Lund, J. W. G., Sanger, J. E., and Dean, W. E., 1974, Some relationships between algal standing crop, water chemistry and sediment chemistry in the English lakes, *Limnol. Oceanogr.* 19:601–617.

Gorlenko, W. M., and Kusnezow, E. D., 1972, Über die photosynthesierenden Bakterien des Kononjer Sees, *Arch. Hydrobiol.* 70:1–13.

Govindjee, and Govindjee, R., 1975, Introduction to photosynthesis, in: *Bioenergetics of Photosynthesis* (Govindjee, ed.), pp. 1–50, Academic Press, New York.

Govindjee, R., Robinovitch, E., and Govindjee, 1968, Maximum quantum yield and action spectrum of photosynthesis and fluorescence in *Chlorella*, *Biochim. Biophys. Acta* 162:539–544.

Greenwood, P. H., 1976, Lake George, Uganda, *Philos. Trans. R. Soc. London Ser. B Biol. Sci.* 274:375–391.

Gromet-Elhanan, Z., 1977, Electron transport and photophosphorylation in photosynthetic bacteria, in: *Encyclopedia of Plant Physiology* (A. Trebst and M. Avron, eds.), Vol. 5, pp. 637–662, Springer-Verlag, Berlin.

Gusev, M. V., 1962, The succession of "bloom" species of blue-green algae and some causal factors, *Dokl. Akad. Nauk SSSR* 147:947–950.

Hamberg, M., Samuelsson, B., Björkhem, I., and Danielsson, H., 1974, Oxygenases in fatty acid and steroid metabolism, in: *Molecular Mechanisms of Oxygen Activation* (O. Hayaishi, ed.), pp. 29–85, Academic Press, New York and London.

Hayaishi, O., 1974, General properties and biological functions of oxygenases, in: *Molecular Mechanisms of Oxygen Activation* (O. Hayaishi, ed.), pp. 1–28, Academic Press, New York and London.

Hinze, G., 1903, Über Schwefeltropfen im Innern von Oscillarien, *Ber. Dtsch. Bot. Ges.* 21:394–398.

Holton, R. W., and Myers, J., 1967a, Water-soluble cytochromes from a blue-green alga. I. Extraction, purification and spectral properties of cytochromes c (549, 552, and 554, *Anacystis nidulans*), *Biochim. Biophys. Acta* 131:362–374.

Holton, R. W., and Myers, J., 1967b, Water soluble cytochromes from a blue-green alga. II. Physiochemical properties and quantitative relationships of cytochromes c (549, 552, and 554, *Anacystis nidulans*), *Biochim. Biophys. Acta* 131:375–384.

Huber-Pestalozzi, G., 1938, Das Phytoplankton des Subwassers. 1, in: *Die Binnengewässer* (A. Thienemann, ed.), Vol. 16, pp. 1–89, E. Schweizerbart'sche Verlagsbuchhandlung, Stuttgart.

Hutchinson, G. E., 1967, *A Treatise on Limnology*, Vol. 2, John Wiley and Sons, New York.

Iltis, A., 1971, Note sur *Oscillatoria* (sous-genre *Spirulina platensis Nordst.*), Bourrelly (Cyanophyta) au Tchad, *Cah. O.R.S.T.O.M. Sér, Hydrobiol.* 5:53–72.

Izawa, S., and Good, N. E., 1972, Inhibition of photosynthetic electron transport and photophosphorylation, in: *Methods in Enzymology* (A. San Pietro, ed.), Vol. 24, pp. 355–377, Academic Press. New York and London.

Jagendorf, A. T., 1975, Mechanisms of photophosphorylation, in: *Bioenergetics of Photosynthesis* (Govindjee, ed.), pp. 414–492, Academic Press, New York.

Jones, L. W., and Myers, J., 1964, Enhancement in the blue-green alga, *Anacystis nidulans*, *Plant Physiol.* 39:938–946.

Jørgensen, B. B., 1977, The sulfur cycle of a coastal marine sediment (Limfjorden, Denmark), *Limnol. Oceanogr.* 22:814–832.

Jørgensen, B. B., and Cohen, Y., 1977, Solar Lake (Sinai). V. The sulfur cycle of the benthic cyanobacterial mats, *Limnol. Oceanogr.* 22:657–666.

Jørgensen, B. B., and Fenchel, T., 1974, The sulfur cycle of a marine sediment model system, *Mar. Biol.* 24:189–201.

Juday, C., 1934, The depth distribution of some aquatic plants, *Ecology* 15:325–326.

Kenyon, C. N., 1972, Fatty acid composition of unicellular strains of blue-green algae, *J. Bacteriol.* 109:827–834.

Kenyon, C. N., Rippka, R., and Stanier, R. Y., 1972, Fatty acid composition and physiological properties of some filamentous blue-green algae, *Arch. Mikrobiol.* 83:216–236.

Kessler, E., 1974, Hydrogenase, photoreduction and anaerobic growth, in: *Algal Physiology and Biochemistry* (W. D. P. Stewart, ed.), pp. 456–473, Blackwell Scientific Publishers, Oxford.

Kjeldsen, C. K., and Walsby, A. E., 1977, Nitrogenase activity in a marine blue-green algal mud flat community, *J. Phycol.* 13: Suppl., p. 36.

Knobloch, K., 1966a, Photosynthetische Sulfidoxidation grüner Pflanzen. I., *Planta* 70:73–78.

Knobloch, K., 1966b, Photosynthetische Sulfidoxidation grüner Pflanzen. II. Wirkung von Stoffwechselinhibitoren, *Planta* 70:172–186.

Knobloch, K., 1969, Sulfide oxidation via photosynthesis in green algae, in: *Progress in Photosynthesis Research* (H. Metzner, ed.), Vol. II, pp. 1032–1034, International Union of Biological Sciences, Tübingen.

Krogmann, D. W., 1973, Photosynthetic reactions and components of thylakoids, in: *The Biology of Blue-Green Algae* (N. G. Carr and B. A. Whitton, eds.), pp. 80–98, Blackwell Scientific Publications, Oxford.

Krogmann, D. W., 1977, Blue-green algae, in: *Encyclopedia of Plant Physiology* (A. Trebst and M. Avron, eds.), Vol. 5, pp. 625–636, Springer-Verlag, Berlin.

Krumbein, W. E., and Cohen, Y., 1977, Primary production, mat formation and lithification: Contribution of oxygenic and facultative anoxygenic cyanobacteria, in: *Fossil Algae* (E. Flügel, ed.), pp. 37–56, Springer-Verlag, Berlin.

Kusai, A., and Yamanaka, T., 1973a, The oxidation mechanisms of thiosulphate and sulphide in *Chlorobium thiosulphatophilum:* Roles of cytochrome c-551 and cytochrome c-553, *Biochim. Biophys. Acta* 325:304–314.

Kusai, A., and Yamanaka, T., 1973b, Cytochrome c (553, *Chlorobium thiosulfatophilum*) is a sulphide cytochrome c reductase, *FEBS Lett.* **34**:235–237.

Lascelles, J., 1959, Adaptation to form bacteriochlorophyll in *Rhodopseudomonas spheroides:* Change in activity of enzymes concerned in pyrrole synthesis, *Biochem. J.* **72**:508–518.

Lewin, J. C., 1950, Obligate autotrophy in *Chlamydomonas moewusii* Gerloff, *Science* **112**:652–653.

Lund, J. W. G., 1959, Buoyancy in relation to the ecology of the freshwater phytoplankton, *Br. Phycol. Bull.* **1**:1–17.

Lund, J. W. G., 1973, Changes in the biomass of blue-green and other algae in an English Lake from 1945–1969, in: *Proceedings of the Symposium on Taxonomy and Biology of Blue-Green Algae* (T. V. Desikachary, ed.), pp. 305–327, University of Madras Press, Madras.

Margulis, L., 1968, Evolutionary criteria in thallophytes: A radical alternative, *Science* **161**:1020–1022.

McCord, J. M., Keele, B. B., Jr., and Fridovich, I., 1971, An enzyme-based theory of obligate anaerobiosis: The physiological function of superoxide dismutase, *Proc. Natl. Acad. Sci. U.S.A.* **68**:1024–1027.

Meffert, M. E., 1975, Analysis of the population dynamics of *Oscillatoria redekei* Van Goor in Lake Edeberg, *Verh. Int. Ver. Limnol.* **19**:2682–2688.

Meffert, M. E., and Krambeck, H. J., 1977, Planktonic blue-green algae of the *Oscillatoria redekei* group, *Arch. Hydrobiol.* **79**:149–171.

Nakamura, H., 1937, Über das Auftreten des Schwefelkügelschens im Zellinnern von einigen niederen Algen, *Bot. Mag.* **51**:529–533.

Nakamura, H., 1938, Über die Kohlensäureassimilation bei niederen Algen in Anwesenheit des Schwefelwasserstoffes, *Acta Phytochim.* **10**:271–281.

Nührenberg, B., Lesemann, D., and Pirson, A., 1968, Zur Frage eines anaeroben Wachstums von einzelligen Grünalgen, *Planta* **79**:162–180.

Odum, H. T., 1967, Biological circuits and the marine systems of Texas, in: *Pollution and Marine Ecology* (T. A. Olson and F. J. Burgess, eds.), pp. 99–157, John Wiley–Interscience, New York.

Ogawa, T., and Vernon, L., 1971, Increased content of cytochromes 554 and 562 in *Anabaena variabilis* cells grown in the presence of diphenylamine, *Biochim. Biophys. Acta* **226**:88–97.

Ohle, W., 1964, Interstitiallösungen der Sedimente, Nährstoffgehalt des Wassers und Primärproduktion des Phytoplanktons in Seen, *Helgol. Wiss. Meeresunters.* **10**:411–429.

Olson, J. M., 1978, Precambrian evolution of photosynthetic and respiratory organisms, in: *Evolutionary Biology* (M. K. Hecht, W. C. Steere, and B. Wallace, eds.), Vol. 11, pp. 1–37, Plenum Press, New York.

Oren, A., and Padan, E., 1978, Induction of anaerobic, photoautotrophic growth in the cyanobacterium, *Oscillatoria limnetica, J. Bacteriol.* **133**:558–563.

Oren, A., Padan, E., and Avron, M., 1977, Quantum yields for oxygenic and anoxygenic photosynthesis in the cyanobacterium *Oscillatoria limnetica, Proc. Natl. Acad. Sci. U.S.A.* **74**:2152–2156.

Padan, E., and Schuldiner, S., 1978, Energy transduction in the photosynthetic membranes of the cyanobacterium (blue-green alga) *Plectonema boryanum, J. Biol. Chem.* **253**:3281–3286.

Padan, E., Zilberstein, D., and Rottenberg, H., 1976, The proton electrochemical gradient in *Escherichia coli* cells, *Eur. J. Biochem.* **63**:533–541.

Parson, W. W., 1974, Bacterial photosynthesis, *Annu. Rev. Microbiol.* **28**:41–59.

Peschek, G. A., 1977, Anoxygenic photosynthetic CO_2 fixation in the prokaryotic alga *Anacystis nidulans*, in: *Abstracts of the 4th International Congress on Photosynthesis, Reading, U.K.* (J. Coombs, ed.), p. 294, UK ISES, 21 Albemarle St., London.

Pfennig, N., 1967, Photosynthetic bacteria, *Annu. Rev. Microbiol.* 21:285–324.

Pfennig, N., 1975, The phototrophic bacteria and their role in the sulfur cycle, *Plant Soil* 43:1–16.

Pfennig, N., 1977, Phototrophic green and purple bacteria: A comparative, systematic survey, *Annu. Rev. Microbiol.* 31:275–290.

Pfennig, N., and Trüper, H. G., 1974, The phototrophic bacteria, in: *Bergey's Manual of Determinative Bacteriology* (S. T. Cowan, J. G. Holt, J. Liston, R. G. E. Murray, C. F. Niven, A. W. Ravin, and R. Y. Stanier, eds.), pp. 24–75, Williams and Wilkins, Baltimore.

Pierson, B. K., and Castenholtz, R. W., 1974, A phototrophic gliding filamentous bacterium of hot springs, *Chloroflexus aurantiacus*, gen. and sp. nov., *Arch. Microbiol.* 100:5–24.

Potts, M., and Whitton, B. A., 1977, Nitrogen fixation by blue-green algal communities in the intertidal zone of the Lagoon of Aldabra Atoll. *Oecologia* 27:275–283.

Potts, M., Krumbein, W. E., and Metzger, J., 1978, Nitrogen fixation rates in anaerobic sediments determined by acetylene reduction, a new ^{15}N field assay, and simultaneous total N and ^{15}N determination, in: *Environmental Biogeochemistry and Geomicrobiology* (W. E. Krumbein, ed.), Part 3, pp. 753–769, Ann Arbor Science Publ., Ann Arbor, Michigan.

Raboy, B., and Padan, E., 1978, Active transport of glucose and α-methylglucoside in the cyanobacterium *Plectonema boryanum*, *J. Biol. Chem.* 253:3287–3291.

Ravera, O., and Vollenweider, R. A., 1968, *Oscillatoria rubescens* D. C. as an indicator of pollution, *Schweiz. Z. Hydrol.* 30:374–380.

Renaut, J., Sasson, A., Pearson, H. W., and Stewart, W. D. P., 1975, Nitrogen-fixing algae in Morocco, in: *Nitrogen Fixation by Free-Living Micro-organisms* (W. D. P. Stewart, ed.), pp. 229–246, Cambridge University Press, Cambridge.

Reynolds, C. S., 1973, Growth and buoyancy of *Microcystis aeruginosa* Kütz. emend. Elenkin in a shallow eutrophic lake, *Proc. R. Soc. London Ser. B Biol. Sci.* 184:29–50.

Reynolds, C. S., 1976, Succession and vertical distribution of phytoplankton in response to thermal stratification in a lowland mere, with special reference to nutrient availability, *J. Ecol.* 64:529–551.

Reynolds, C. S., and Rogers, D. A., 1976, Seasonal variations in the vertical distribution and buoyancy of *Microcystis aeruginosa* Kütz. emend. Elenkin in Rostherne Mere, England, *Hydrobiologia* 48:17–26.

Reynolds, C. S., and Walsby, A. E., 1975, Water-blooms, *Biol. Rev.* 50:437–481.

Rich, F., 1932, Reports on the Percy Sladen Expedition to some Rift Valley lakes in Kenya in 1929, IV. Phytoplankton from the Rift Valley lakes in Kenya, *Ann. Mag. Nat. Hist. Ser.* 10(10):233–262.

Rippka, R., 1972, Photoheterotrophy and chemoheterotrophy among unicellular blue-green algae, *Arch. Mikrobiol.* 87:93–98.

Rippka, R., and Waterbury, J. B., 1977, The synthesis of nitrogenase by non-heterocystous cyanobacteria, *FEMS Microbiol. Lett.* 2:83–86.

Robarts, R. D., and Southall, G. C., 1977, Nutrient limitation of phytoplankton growth in seven tropical man-made lakes with special reference to Lake McIlwaine, Rhodesia, *Arch. Hydrobiol.* 79:1–35.

Roy, A. B., and Trudinger, P. A., 1970, *The Biochemistry of Inorganic Compounds of Sulphur*, Cambridge University Press, Cambridge.

Schindler, D. W., and Holmgren, S. K., 1971, Primary production and phytoplankton in the experimental Lakes Area, north-western Ontario, and other low-carbonate waters, and a liquid scintillation method for determining ^{14}C activity in photosynthesis, *J. Fish. Res. Board Can.* **28**:189–201.

Schopf, J. W., 1974, Paleobiology of the Precambrian: The age of blue-green algae, in: *Evolutionary Biology* (T. Dobzhansky, M. K. Hecht, and W. C. Steere, eds.), Vol. 7, pp. 1–43, Plenum Press, New York.

Schuster, W. H., 1949, *De Viscultuur in de Kustvijvers op Java*, Dept. Landbw. en Viss., Publ. 2.

Schwabe, G. H., 1960, Über den thermobionten Kosmopoliten *Mastigocladus laminosus* Cohn. Blaualgen und Lebensraum, *Schweiz. Z. Hydrol.* **22**:757–792.

Serruya, C., 1972, Metalimnic layer in Lake Kinneret, Israel, *Hydrobiologia* **40**:355–359.

Serruya, C., Edelstein, M., Pollingher, U., and Serruya, S., 1974, Lake Kinneret sediments: Nutrient composition of the pore water and mud-water exchanges, *Limnol. Oceanogr.* **19**:489–508.

Serruya, S., 1975, Wind, water temperature and motions in Lake Kinneret: General pattern, *Verh. Int. Ver. Limnol.* **19**:73–87.

Setlike, I., 1957, Light–dark transients in oxygen exchange of blue-green algae, *Biochim. Biophys. Acta* **24**:436–437.

Sheridan, R. P., 1973, Hydrogen sulfide production by *Synechococcus lividus Y52-s, J. Phycol.* **9**:437–445.

Shilo, M., 1972, Toxigenic algae, in: *Progress in Industrial Microbiology* (D. J. D. Hockenhull, ed.), Vol. 11, pp. 233–265, Churchill Livingstone, Edinburgh and London.

Simonis, M., and Urbach, W., 1973, Photophosphorylation in vivo, *Annu. Rev. Plant Physiol.* **24**:89–114.

Singh, R. N., 1955, Limnological relations of Indian inland waters with special reference to water blooms, *Verh. Int. Ver. Theor. Angew. Limnol.* **12**:831–836.

Singh, R. N., 1961, *The Role of Blue-Green Algae in Nitrogen Economy of Indian Agriculture*, Indian Council for Agricultural Research, New Delhi.

Sirenko, L. A., 1972, *Fiziologicheskie Osnovy Razmnozheniya, Sine-zelenykh Vodoroslei v Vodokhranilishchakh*, Kiev, pp. 88–162 (available as National Lending Library translation RTS 8132).

Sirenko, L. A., Chernousova, V. M., Arendarchuk, V. V., and Kozitskaya, V. N., 1969, Factors of mass development of blue-green algae, *Hydrobiol. J.* **5**:1–8.

Smith, A. J., and Hoare, D. S., 1977, Specialist phototrophs, litotrophs, and methylotrophs: Unity among a diversity of prokaryotes? *Bacteriol. Rev.* **41**:419–448.

Smith, A. J., London, J., and Stanier, R. Y., 1967, Biochemical basis of obligate autotrophy in blue-green algae and thiobacilli, *J. Bacteriol.* **94**:972–983.

Sorokin, J. U. I., 1965, On the trophic role of chemosynthesis and bacterial biosynthesis in waterbodies, in: *Primary Production in Aquatic Environments* (C. R. Goldman, ed.), pp. 187–205, Memorie dell'Istituto Italiano di Idrobiologia, 18th Suppl., University of California Press, Berkeley.

Sournia, A., 1976, Ecologie et productivité d'une Cyanophycée en milieu corallien: *Oscillatoria limnosa* Agardh, *Phycologia* **15**:363–366.

Stanier, R. Y., 1974, The origins of photosynthesis in eukaryotes, in: *Evolution in the Microbial World*, Twenty-Fourth Symposium of the Society for General Microbiology (M. J. Carlile and J. J. Skehel, eds.), pp. 219–240, Cambridge University Press, Cambridge.

Stanier, R. Y., and Cohen-Bazire, G., 1977, Phototrophic prokaryotes: The cyanobacteria, *Annu. Rev. Microbiol.* **31**:225–274.

Stanier, R. Y., Kunisawa, R., Mandel, M., and Cohen-Bazire, G., 1971, Purification and properties of unicellular blue-green algae (order Chroococcales), *Bacteriol. Rev.* **35**:171–205.

Steemann-Nielsen, E., 1975, *Marine Photosynthesis with Special Emphasis on the Ecological Aspects*, Elsevier Amsterdam.

Stepánek, M., and Chalupa, J., 1958, Limnological study of the Reservor Sedlice, near Zeliv. II. Biological Part, Praha: *Sb. Vys. Sk. Chem. Technol. Praze.*

Stewart, W. D. P., 1973, Nitrogen fixation, in: *The Biology of Blue-Green Algae* (N. G. Carr and B. A. Whitton, eds.), pp. 260–278, Blackwell Scientific Publications, Oxford.

Stewart, W. D. P., and Lex, M., 1970, Nitrogenase activity in the blue-green alga *Plectonema boryanum* Strain 594, *Arch. Microbiol.* 73:250–260.

Stewart, W. D. P., and Pearson, H. W., 1970, Effects of aerobic and anaerobic conditions on growth and metabolism of blue-green algae, *Proc. R. Soc. Lond. Ser. B Biol. Sci.* 175:293–311.

Strickland, J. D. H., 1965, Production of organic matter in the primary stages of the marine food chain, in: *Chemical Oceanography* (J. P. Riley and G. Skirrow, eds.), pp. 478–610, Academic Press, London and New York.

Susor, W. A., and Krogmann, D. W., 1966, Triphosphopyridine nucleotide photoreduction by cell-free preparations of *Anabaena variabilis*, *Biochim. Biophys. Acta* 120:65–72.

Talling, J. F., Wood, R. B., Prosser, M. V., and Baxter, R. M., 1973, The upper limit of photosynthetic productivity of phytoplankton: Evidence from Ethiopian soda lakes, *Freshwater Biol.* 3:53–76.

Taylor, B. F., Lee, C. C., and Bunt, J. S., 1973, Nitrogen fixation associated with the marine blue-green alga *Trichodesmium* as measured by the acetylene-reduction technique, *Arch. Mikrobiol.* 88:205–212.

Tel-Or, E., Luijk, L. W., and Packer, L., 1977, An inducible hydrogenase in cyanobacteria enhances N_2 fixation, *FEBS Lett.* 78:49–52.

Thomas, E. A., 1949, Sprungschichtneigung im Zürichsee durch Sturm, *Schweiz. Z. Hydrol.* 11:527–545.

Thomas, E. A., 1950, Auffällige biologische Folgen von Sprungschichtneigungen im Zürichsee, *Schweiz. Z. Hydrol.* 12:1–24.

Topachevskii, A. V., Braginskii, L. P., and Sirenko, J. A., 1969, Massive development of blue-green algae as a product of the ecosystem of a reservoir, *Hydrobiol. J.* 5:1–10.

Trebst, A., 1974, Energy conservation in photosynthetic electron transport of chloroplasts, *Annu. Rev. Plant Physiol.* 25:423–458.

Trüper, H. G., 1973, The present state of knowledge of sulfur metabolism in phototrophic bacteria, in: *Abstracts of Symposium on Prokaryotic Photosynthetic Organisms, Freiburg, Germany, September, 1973* (organized by G. Drews, N. Pfennig, and R. Y. Stanier), pp. 160–166, Deutsche Forschungsgemeinschaft, International Union of Biological Sciences, Freiburg, Germany.

Utkilen, H. C., 1975, Thiosulfate as electron donor in the blue-green alga *Anacystis nidulans*, *J. Gen. Microbiol.* 95:177–180.

Utermöhl, H., 1925, Limnologische Phytoplanktonstudien. Die Besiedlung ostholsteinischer Seen mit Schwebpflanzen, *Arch. Hydrobiol. Suppl.* 5:1–527.

Van Niel, C. B., 1931, On the morphology and physiology of the purple and green sulfur bacteria, *Arch. Mikrobiol.* 3:1–112.

Van Niel, C. B., 1963, A brief survey of the photosynthetic bacteria, in: *Bacterial Photosynthesis* (H. Gest, A. San Pietro, and L. P. Vernon, eds.), pp. 459–467, Antioch Press, Yellow Springs, Ohio.

Venkataraman, G. S., 1975, The role of blue-green algae in tropical rice cultivation, in: *Nitrogen Fixation by Free-Living Micro-organisms* (W. D. P. Stewart, ed.), pp. 207–218, Cambridge University Press, Cambridge.

Viner, A. B., 1975, The sediment of Lake George (Uganda). I. Redox potentials oxygen consumption and carbon dioxide output, *Arch. Hydrobiol.* 76:181–197.

Viner, A. B., and Smith, I. R., 1973, Geographical, historical and physical aspects of Lake George, *Proc. R. Soc. London Ser. B Biol. Sci.* **184**:235–270.

Walsby, A. E., 1970, The nuisance algae: Curiosities in the biology of planktonic blue-green algae, *Water Treat. Exam.* **19**:359–373.

Walsby, A. E., 1975, Gas vesicles, *Annu. Rev. Plant Physiol.* **36**:427–439.

Walsby, A. E., and Klemer, A. R., 1974, The role of gas-vacuoles in the microstratification of a population of *Oscillatoria agardhii* var. *isothrix* in Deming Lake, Minnesota, *Arch. Hydrobiol.* **74**:375–392.

Weissman, J. C., and Benemann, J. R., 1977, Hydrogen production by nitrogen starved cultures of *Anabaena cylindrica*, *Appl. Environ. Microbiol.* **33**:123–131.

Weller, D., Doemel, W., and Brock, T. D., 1975, Requirement of low oxidation–reduction potential for photosynthesis in a blue-green alga (*Phormidium* sp.), *Arch. Microbiol.* **104**:7–13.

Wetzel, R. G., 1975, *Limnology*, pp. 287–354, W. B. Saunders, Philadelphia.

Whitton, B. A., and Sinclair, C., 1975, Ecology of blue-green algae, *Sci. Prog. (Oxford)* **62**:429–446.

Wohler, R. J., and Hartmann, R. T., 1973, Some characteristics of an *Oscillatoria*-dominated metalimnetic phytoplankton community, *Ohio J. Sci.* **73**:297–306.

Wood, B. J. B., 1974, Fatty acids and saponifiable lipids, in: *Algal Physiology and Biochemistry* (W. D. P. Stewart, ed.), pp. 236–265, Blackwell Scientific Publications, Oxford.

Wood, E. J. F., 1965, *Marine Microbial Ecology*, Chapman and Hall, London; Reinhold, New York.

Wundsch, H. H., 1940, Beiträge zur Fischereibiologie märkischer Seen. VI. Die Entwicklung eines besonderen Seentypus (H$_2$S-*Oscillatorien*-Seen) im Fluss–Seengebiet der Spree und Havel und seine Bedeutung für die fischereibiologischen Bedingungen in dieser Region, *Z. Fisch.* **38**:444–658.

Zimmerman, U., 1969, Ökologische und physiologische Untersuchungen an der planktonischen Blaualge *Oscillatoria rubescens* D. C. unter besonderer Berücksichtigung von Licht und Temperatur, *Schweiz. Z. Hydrol.* **31**:1–58.

2

The Rumen Fermentation: A Model for Microbial Interactions in Anaerobic Ecosystems

MEYER J. WOLIN

1. Introduction

A ruminant can be thought of as a fermentation factory (Fig. 1). The animal ingests plant polymers (in grasses, hay, corn, silage, etc.) which are the raw material for its fermentation. Preliminary processing occurs in the oral cavity and consists mainly of comminution of food by mastication. The plant material is then swallowed and transported to the ruminant's complex stomach. The stomach, called the rumen reticulum or, more simply, rumen, is the site of fermentation. A massive community of microorganisms, bacteria and protozoa, ferments the plant material to short-chain volatile fatty acids, methane, and carbon dioxide. The acids are removed from the rumen by absorption into the bloodstream and are subsequently used as the animal's primary sources of energy and carbon. Gases are waste products and are removed by belching. The fermentation provides nutrients and energy for the growth and division of the microbial populations participating in the fermentation. Microorganisms and undigested food are semicontinuously removed from the rumen by passage to the lower part of the ruminant's digestive tract. Digestive processes then occur that are similar to those of monogastric animals and include gastric digestion in a terminal compartment of the complex stomach called the abomasum

MEYER J. WOLIN • Division of Laboratories and Research, New York State Department of Health, Albany, New York 12201, U.S.A.

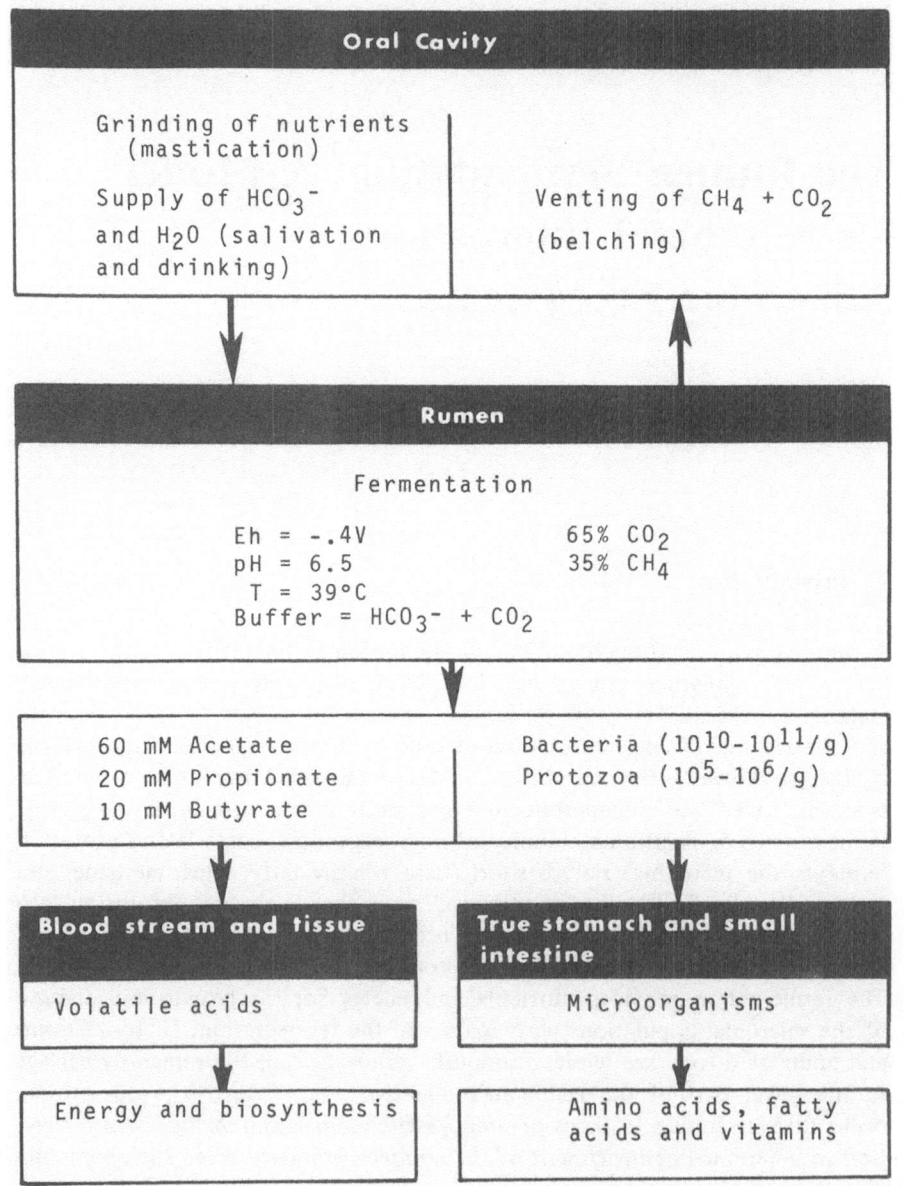

Figure 1. Schematic representation of the ruminant as a factory for the conversion of food to animal products.

and digestion in the small intestine. Digestion of the microbial mass exiting from the rumen provides the animal with its major source of amino acids and water-soluble vitamins. Undigested microbial and feed residues and microorganisms that inhabit the large intestine leave the animal in the feces.

This review will be concerned with the central feature of the factory, the fermentation in the rumen and the characteristics of the rumen microbial ecosystem. The environment of the rumen will be described to emphasize the parameters that are important in its function as a finely controlled, semicontinuous fermentation apparatus. Major characteristics of the microbial community will be outlined. Food-chain relations between the rumen microbial community and the animal and between different microbial populations in the rumen will be reviewed. A major subject that will be emphasized is how the populations in the rumen interact to produce its characteristic fermentation products, the volatile fatty acids, methane, and carbon dioxide. Since the general characteristics of the rumen are similar to those of other important microbial ecosystems—e.g., the large intestine of monogastric animals, anaerobic sediments, and waste fermentation systems—similarities and differences between these systems will be discussed. Many important and interesting details of ruminant anatomy, physiology, and nutrition and characteristics of the rumen microbial community will not be reviewed because the major goal is to present the general principles that govern the characteristics of the rumen ecosystem. The reader is referred to several excellent reviews which present detailed information about the omitted subjects (Bryant, 1959; Hungate et al., 1964; Hungate, 1966; Janis, 1976; Bryant, 1977). Figure 1 is a flow diagram of the ruminant factory that can be used as a guide to the following discussion of the ecosystem.

2. The Rumen Environment

The rumen, as a semicontinuous fermentation apparatus, has several features that distinguish it from conventional man-made fermentation equipment. The inner epithelial wall of the rumen is semipermeable and selectively transports small molecules to and from the blood of the animal. For example, volatile fatty acids are transported from the rumen to the blood, and bicarbonate ions are transported in the reverse direction. Neural responses to specific stimuli activate physiological mechanisms for eructation (to release fermentation gases), activate contractions to mix and move rumen contents, and activate the mechanism for regurgitation of solids. The regurgitated solids are reground in the oral cavity ("chewing the cud") to permit continued comminution of solid substrates even after they enter the fermentation chamber and fermentation is initiated. The biological features of the rumen provide elegant mechanisms for fermentation of solid substrates, product removal, maintenance of pH, and disposal of fermentation gases.

Although the rumen is a biological fermentation unit, it, like artificial fermentation units, operates under well-defined environmental conditions that are extremely important regulators of the kinds, amounts, and physiological activities of the microorganisms in the ecosystem. The rumen environment is anaerobic. The gas phase is essentially all carbon dioxide and methane, although trace amounts of other gases (H_2, N_2, and O_2) have been reported. The gas composition is approximately 65% carbon dioxide and 35% methane. Methane arises strictly as a fermentation product. Carbon dioxide is produced both from fermentation and by neutralization of acids by bicarbonate ions entering the rumen from the saliva and blood. The amounts of gases produced from the fermentation are very large and are sufficient to significantly dilute ingested gases. A 500-kg bovine eructates approximately 200 liters of methane per day. Small amounts of air that enter the rumen during swallowing do not, therefore, significantly alter the composition of the gas phase. Any small amounts of O_2 entering the system are quickly used up by oxidative metabolism by rumen microorganisms. This oxidative activity represents only a minute portion of the metabolism of the tremendous amount of organic matter entering the ecosystem. In this sense, the rumen is like a highly eutrophic aquatic environment where limiting O_2 can support the oxidation of only an insignificantly small amount of the organic matter in the system. The oxidation–reduction potential of the rumen is low and is approximately -350 mV (E_0' at pH 6.5), which is close to the oxidation–reduction potential of the hydrogen electrode at the same pH.

The pH of the rumen is approximately 6.5. Neutral plant polymers are fermented to predominantly acid products, and the pH would drop significantly during fermentation if it were not for the strong bicarbonate–carbon dioxide buffering system of the rumen. Salivary bicarbonate is the major neutralizing agent for the acids produced during fermentation. The temperature of the rumen is maintained at approximately 39°C through physiological regulation by the animal.

It is important to emphasize the semicontinuous nature of the fermentation system. A 500-kg bovine has approximately 70 liters of contents in its rumen. During a single day, approximately 100 liters of liquid enters the fermentation vat (about 60 liters of saliva and about 40 liters of drinking water). In order to maintain a volume of 70 liters, 100 liters has to be removed from the rumen each day. This would be equivalent to a dilution rate of 1.4 per day (or a turnover time of 0.7 days) for a truly continuous culture. The system, however, is not truly continuous. Feeding of nutrients is sporadic, as is the inflow of liquid, and the nutrient suspension is not homogeneous. In addition, mixing in the rumen is imperfect, and the regurgitation of ingested material adds to the discontinuity of the system. Solids have a turnover time that is approximately three times longer than liquids.

The environmental parameters set some obvious constraints on the populations of microorganisms one might expect to find in the ecosystem. Obligate aerobes would not be expected to occur in significant concentrations. Dominant species have to be either facultative anaerobes or anaerobes. The dilution rate sets some rough upper limits on the growth rate constants or generation times of organisms that can be maintained in the ecosystem. Some of the microorganisms are associated with solids, and it can be roughly estimated that associated organisms with generation times of about two days or less would not be washed out of the ecosystem. The protozoa of the rumen have generation times of the order of a day or two and probably would not be able to remain in the rumen if it was a homogeneously mixed reactor with a turnover time of 0.7 days. Most of the bacteria in the system have much shorter generation times than 0.7 days. The methane-forming bacteria do not, and they remain in the system because the dilution rate can accommodate their long generation times. On the other hand, much longer turnover times, i.e., five to ten days, would probably encourage the development of bacteria that can participate in the complete bioconversion of organic carbon to methane and carbon dioxide as occurs in waste disposal systems. This would be detrimental to the animal because it would no longer have the volatile fatty acids available that it needs as sources of carbon and energy in its metabolism. The elemental carbon, hydrogen, and oxygen of the acids would end up instead as carbon, hydrogen, and oxygen of methane and carbon dioxide.

Temperature, oxidation–reduction potential, and pH are constraints on the types of microorganisms that can exist in the rumen only in the sense that microorganisms would not be present that have obligatory requirements for conditions that are significantly different from those found in the rumen. It is not unreasonable to argue, however, that evolution of the system has selected against organisms that have ranges of activity that include but go far beyond the normal environmental ranges of the ecosystem. For example, O_2-using facultative anaerobes have genetic information and enzymes for metabolic activities they would never use in the rumen environment. Their presence in significant concentrations would have been selected against in favor of organisms that do not have any significant ability to use O_2. Similarly, it would be expected that evolution has selected for organisms with optimal growth temperatures close to that of the environment and with maximal and minimal temperatures not far from the temperature of the rumen. In general, the physiological characteristics of microorganisms reflect the evolution of the most efficient physiological mechanisms for growth and survival in the natural habitat.

The rumen environment is described in detail in Hungate (1966).

3. Rumen Bacteria and Protozoa

Almost all the significant rumen bacteria are nonsporing anaerobes. A few species of facultative anaerobes are present that have no significant ability to use O_2. Occasionally, significant concentrations of spore-forming anaerobes have been detected in the rumen.

A few words are necessary to clarify what constitutes a significant bacterial species (Bryant, 1959). The rumen is an open system. Any microorganisms that enters or exists in the oral cavity can enter the rumen. Most will not grow or will grow very poorly in the rumen, but as long as they are not killed they can be isolated from the rumen by selective procedures. Criteria have to be established to separate transients and those that barely manage to eke out an existence in the community from the dominant, important species. Since total counts of bacteria are about 1×10^{10} to 5×10^{10} per gram of rumen contents, it is common practice to consider species whose concentrations are 10^8-10^{10} per gram as members of the dominant community. It has always been the case that these species demonstrate physiological and nutritional characteristics that are consistent with the characteristics of the ecosystem. There is always the possiblity of neglecting large sized, less numerous species which may significantly contribute to bacterial mass and activity, or organisms in low concentration which may have high activity for some critical ecosystem function. Practically speaking, however, it has been possible to associate most of the important rumen activities with species that are present in high concentrations. Genera and species of protozoa can be identified and enumerated by microscopy. This provides a more direct method for evaluating the significance of specific taxa than is possible with most of the bacteria.

Table I shows approximations of the mass of bacteria and protozoa in the rumen and estimates of the amount of protein they contribute to rumen contents. Rumen contents represent approximately 15% of the live weight of a bovine and contain about 65% of the weight of food ingested by the animal. Table II shows the distribution of material in the intestinal tract of a cow as percentages of body weight and daily intake of food. Rumen contents are approximately 10–18% dry matter (Hungate, 1975). Microbial cells constitute about 5% of the dry matter, and they represent about 1% of the dry weight of the entire animal (Hungate, 1966; Bryant, 1977). The microbial mass is usually equally distributed between bacteria and protozoa (Table I). Although ruminants can be reared without protozoa by the use of various methods to defaunate or to prevent the development of the fauna (Becker, 1932; Eadie, 1962), the typical rumen is a faunated system. Protozoa are important to, although not essential for, the function of the rumen. Because metabolic activity is significantly greater per cell as cell volume decreases, the contribution of bacteria to the overall fermentation in the rumen is significantly greater than that of the protozoa.

Table I. Dry Weight (g) of Bacteria and Protozoa and
Microbial Protein in the Rumen[a]

	Ruminant	
Microorganism	Bovine[b]	Ovine[c]
Protozoa		
Cells	315	22.5
Cell protein	172	12.3
Bacteria		
Cells	399	28.5
Cell protein	217	15.5

[a] Values are based on data in Hungate et al. (1971). Total bacteria were 0.57 g (dry weight)/
100 ml (3.1 X 10^{10} cells/ml). Total protozoa were 0.45 g (dry weight)/ 100 ml (2.4 X 10^5
cells/ml). Protein is estimated as 54.5% of the dry weight (Reichl and Baldwin, 1975).
[b] Based on 70 liters of rumen contents.
[c] Based on 5 liters of rumen contents.

Most of the important species of rumen bacteria are listed in Table III with their major fermentation products and their major energy sources. The important cellulolytic species are *Ruminococcus albus*, *R. flavefaciens*, *Bacteroides succinogenes*, and possibly *Butyrivibrio fibrisolvens*. Important starch-digesting species include *B. fibrisolvens*, *Selenomonas ruminantium*, *Bacteroides amylophilus*, *Streptococcus bovis*, and *Succinomonas amylolytica*.

The protozoa of the rumen are anaerobic ciliates. There are two major groups, the holotrichs and the entodiniomorphs (also called oligotrichs). The holotrichs have cilia all over the cell surface. They are represented by the genera *Dasytricha* and *Isotricha*. Both ferment soluble sugars, and *Isotricha* ferments starch. Entodiniomorphs have their cilia arranged as complex organelles. One genus, *Diplodinium*, ferments cellulose and starch, and other genera ferment starch, hemicellulose, and pectin. The entodiniomorphs have only a very limited

Table II. Contents of Digestive Organs of a Milk Cow[a]

Organ	Wet wt.		Dry wt.	
	kg	% wt. of animal	kg	% Daily food intake
Rumen reticulum	81.8	15.4	8.56	64.8
Omasum	11.0	2.1	0.70	5.3
Abomasum	3.6	0.7	0.27	2.0
Small intestine	9.4	1.8	0.27	2.0
Large intestine	8.6	1.6	0.38	2.9

[a] From Hungate (1966). The weight of the cow was 533 kg, and the daily food intake was
13.2 kg (dry wt.) of hay.

Table III. Fermentation Products and Energy Sources of Important Rumen Bacteria[a]

Major fermentation products[b]	Species	Energy sources[c]					
		Cellulose	Starch	Xylan	Glucose	Lactate	Glycerol
A, S	*Bacteroides succinogenes*	+	∓	−	+	−	−
A, S, F	*B. amylophilus*	−	+	−	−	−	−
A, S, F	*B. ruminicola*	−	±	∓	−	−	−
A, S, F, H	*Ruminococcus flavefaciens*	±	−	±	±	−	−
A, S	*Succinivibrio dextrinosolvens*	−	−	−	+	−	−
S	*Succinomonas amylolytica*	−	+	−	+	−	−
A, S	*Treponema* sp.	−	−	−	+	−	−
A, E, F, H	*Ruminococcus albus*	±	−	±	±	−	−
A, E, F, L, H	*Lachnospira multiparus*	−	∓	+	−	−	−
B, F, H	*Butyrivibrio fibrisolvens*	∓	±	±	+	−	−
B, F, L	*Eubacterium ruminantium*	−	−	±	+	−	−
A, P, B, H	*Megasphaera elsdenii*	−	−	−	+	+	∓
A, P, L	*Selenomonas ruminantium*	−	∓	−	+	∓	∓
A, P, H	*Veillonella alcalescens*	−	−	−	−	+	−
L	*Streptococcus bovis*	−	+	−	+	−	−
L	*Lactobacillus vitulinus*	−	−	−	+	−	−
CH₄	*Methanobacterium ruminantium*	H₂ and CO₂, formate					

[a] Adapted from Bryant (1977).

[b] Abbreviations: A, acetate; S, succinate; F, formate; H, hydrogen; E, ethanol; B, butyrate; P, propionate; L, lactate. CO_2 is produced by many and used by a few species during fermentation.

[c] +, Used by all strains; ∓, used by few strains; −, not used by any strains.

ability to use soluble carbohydrates. Major products of the protozoan fermentation are acetate, butyrate, lactate, H_2, and carbon dioxide. Propionate may be a minor product. Formation of ethanol and succinate has not been reported.

Detailed information about rumen bacteria and protozoa can be found in Bryant (1959), Hungate et al. (1964), and Hungate (1966).

4. Interactions between Microorganisms and the Animal

The ruminant provides nutrients that support the growth of rumen microorganisms. All of the necessary carbon, nitrogen, phosphorus, sulfur, and trace elements are present in the animal's food. The ruminant also contributes substantially to the maintenance of appropriate physical and chemical conditions for its fermentation factory; e.g., it contributes to temperature and pH control, and it controls the dynamics of turnover of the contents of the rumen.

In return for the provision of this excellent habitat, rumen microorganisms provide activities and products that are essential for the animal. First and foremost of these are the microbial cellulases. Cellulose is the most important source of carbon and energy in the ruminant's diet, but the animal itself does not produce enzymes for hydrolyzing cellulose. As the microorganisms use cellulose and other plant carbohydrates as their sources of carbon and energy, they produce large amounts of acetic, propionic, and butyric acids which the animal uses as its sources of carbon and energy. Table IV shows estimates of the amounts of these acid products produced per day in the ovine and bovine. Very large

Table IV. Daily Production of Volatile Fatty Acids
in the Rumen[a]

| | Ruminant | |
Volatile acid	Bovine[b]	Ovine[c]
Acetic		
mol per rumen	62	4.4
kg per rumen	3.7	0.26
Propionic		
mol per rumen	15	1.1
kg per rumen	1.1	0.08
Butyric		
mol per rumen	7	0.5
kg per rumen	0.6	0.04

[a] Values are estimates based on rates of production of 3.7, 0.9, and 0.4 mmol/100 g per hr for acetic, propionic, and butyric acid, respectively (Hungate et al., 1971).
[b] Based on 70 kg of rumen contents.
[c] Based on 5 kg of rumen contents.

amounts become available to the animal from the microbial digestion of plant matter.

It is important to recognize that almost all organic plant constituents that enter the rumen will undergo microbial fermentation whether or not the animal itself has the enzymatic ability to digest them. For example, the ruminant does produce pancreatic amylase which is secreted into the small intestine, but the only starch available to it is that which escapes microbial digestion and the starch-like polysaccharides stored within microbial cells. Proteins entering the rumen are also rapidly hydrolyzed and fermented. The animal's proteases are used mainly to hydrolyze the microbial protein of the rumen after it enters the stomach-like portion of the rumen, the abomasum, and the small intestine.

Discussions of the interrelationships between the microbial community and the animal can be confusing if the ruminants being considered are beef cattle fed using modern methods for the rapid production of tender beef. These animals are often fed large amounts of starch rather than cellulose. If starch and high-quality proteins are available for feeding, the husbandryman is often frustrated by the fact that whatever is fed ends up as whatever the microorganisms produce. There are certain economic scenarios where it is easy to imagine that it would be more beneficial to bypass the microbial community with feeds the animal can directly metabolize because the feed ingredients contain more energy and higher-quality protein than the microbial products. These are, however, very recent and very practical scenarios. One only has to consider a cow grazing in a pasture or a mountain goat nibbling on brush to realize that the rumen system evolved to permit the utilization of plants that many animals cannot digest. Celluloses and hemicelluloses are the main sources of carbon and energy for the ruminants except for the specialized feeding now used to prepare beef cattle for market. The animal generally benefits from the upgrading of protein quality that occurs when poor-quality plant proteins are fermented in the rumen and microbial protein is synthesized and is the major protein subjected to intestinal digestion (Smith, 1975). This upgrading of quality is actually taken advantage of in modern feeding where it is sometimes economically advantageous to feed chemically synthesized urea. The urea is rapidly broken down to ammonia and carbon dioxide by rumen microorganisms, and the ammonia is used as a source of nitrogen for the synthesis of microbial protein (Smith, 1975). A ruminant, therefore, can be maintained on a diet that contains a significant amount of the animal's nitrogen requirement in the form of urea.

In addition to providing protein and a method for using cellulose and hemicellulose, the microbial community synthesizes all of the B vitamins required by the animal. These are released when the microorganisms are digested by the animal. The vitamins are also in high concentration in the fluid of the rumen. Ruminants can suffer indirect B-vitamin deficiencies if vitamin synthesis does not take place in the rumen. When animals are reared on crops raised on cobalt-deficient soils, there is insufficient cobalt for microbial synthesis of vitamin B_{12}, and the animal suffers a vitamin-B_{12}-deficiency disease (Hungate, 1966).

Long-chain fatty acids of microbial lipids are also used by the animal intestinal digestion of microbial cells. The fatty acids are directly incorporated into the lipids of the animal. The origin of the fatty acids of the ruminant's lipid can be identified because of the unique structures of the microbial fatty acids. Some microbial populations also reduce dietary unsaturated to saturated fatty acids, which are then incorporated into ruminant fats.

The interrelationships discussed above probably apply to many animals other than domesticated ruminants that have digestive systems that include a forestomach fermentation of plant matter. Some examples are the camel, the kangaroo, the hippopotamus, and a few species of herbivorous monkeys (Bauchop, 1971; Janis, 1976). The microbial communities of these animals have not been studied in detail, but there is good reason to believe that the general principles governing the operation of these ecosystems are similar to those of domestic ruminants. There are also many herbivores which use a typical monogastric digestion system prior to the digestion of celluloses and hemicelluloses in a portion of the large intestine (e.g., rodents and horses). These herbivores use primarily microbial fermentation for digestion of cellulose and hemicellulose in the posterior portion of the digestive system. Short-chain volatile fatty acids and methane are formed. Although it would be expected that there will be some similarities between the large-intestine fermentation systems and the rumen, it would also be expected that there will be many differences because of the difference in the dynamics of nutrient passage, the more solid nature of contents of the large intestine, and the lack of salivary ingredients. Unfortunately, the large-intestine ecosystems have not been studied to an extent that permits extensive comparisons with the rumen system.

Additional information about the interactions between rumen microorganisms and the animal can be found in Hungate (1966).

5. Nutritional Interactions between Microbial Populations

Nutritional interrelationships between animal and microorganims are of fundamental importance to the maintenance of the ecosystem. There are also important interrelationships between the microbial populations of the rumen which are essential for maintaining the indigenous flora and fauna. There are classical food-chain relationships where the product of digestion by one species serves as a nutrient for another species. Food chains involving carbon and energy sources will be discussed below when fermentation interactions are considered. Nitrogen utilization in the rumen is to some extent related to fermentation but will be discussed here because fermentation of nitrogen compounds produces only a small contribution to products as compared to the fermentation of carbohydrates.

The metabolism of protein in the rumen is summarized in Fig. 2. Proteins entering the rumen are digested and fermented with the release of nitrogen as ammonia (Smith, 1975). Fermentation is rapid and complete. Amino acids do not accumulate in significant concentrations in the environment (Wright and Hungate, 1967). Consequently, the most available form of soluble nitrogen for microbial biosynthesis is ammonia rather than any form of organic nitrogen. Almost all of the major species of rumen bacteria can use ammonia as the major precursor of the cellular nitrogen compounds (Bryant and Robinson, 1962; Bryant, 1974). In fact, several important species use only ammonia and are incapable of growth with amino acids, presumably because they are impermeable to amino acids (Allison, 1969).

Bacteroides ruminicola uses either peptides or ammonia but not amino acids as a source of nitrogen (Pittman *et al.*, 1967). The peptides have to contain between 4 and 20 amino acids. Experiments with rumen contents indicate that proteins are hydrolyzed to peptides that are small enough to enter bacterial cells, where they are used as nitrogen sources and also catabolized to yield ammonia without accumulation of consequential amounts of extracellular amino acids (Wright, 1967). The nitrogen and carbon of amino acids added to rumen contents are catabolized and are inefficiently incorporated into bacterial protein (Wright, 1967). The ability of some of the major species of rumen bacteria to use ammonia and amino acids as major sources of nitrogen for growth is summarized in Table V.

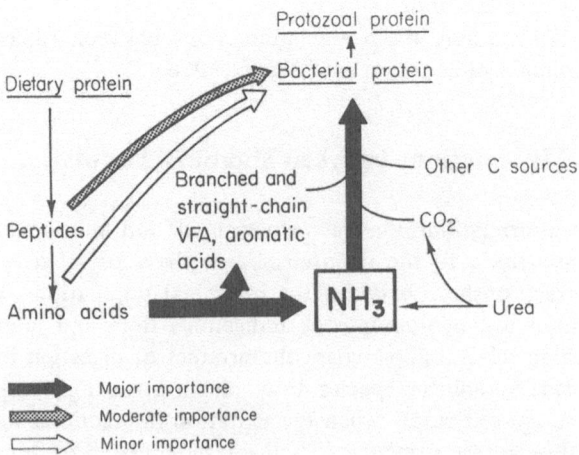

Figure 2. Metabolism of protein in the rumen. VFA, Volatile fatty acids.

Table V. Use of Ammonia and Amino Acids as Major Sources of
Nitrogen by Rumen Bacteria

| | Nitrogen source[a] | |
Species	Ammonia	Amino acids
Bacteroides succinogenes	+	−
B. amylophilus	+	−
B. ruminicola	+	−
Ruminococcus flavefaciens	+	−
R. albus	+	−
Butyrivibrio fibrisolvens	+	+
Megasphaera elsdenii	−	+
Selenomonas ruminantium	+	+
Streptococcus bovis	+	+
Methanobacterium ruminantium	+	−

[a]+, Used by all or most strains; −, not used.

Another habitat-oriented nutritional requirement of at least some rumen
bacteria is associated with the catabolism of proteins and amino acids. The carbon skeletons of the amino acids are fermented to derivative acids, fatty acids
from aliphatic amino acids and aromatic acids from aromatic amino acids. (el-
Shazly, 1952; Scott *et al.*, 1964). The products of fermentation of branched-
chain amino acids are branched-chain fatty acids with one carbon less than the
parent amino acids. Isobutyric acid is derived from Val, isovaleric acid from Ile,
and 2-methylbutyric acid from Leu. Several bacterial populations have nutritional requirements for the branched-chain fatty acids (Bryant, 1974). In some
cases, they are used by the organisms for the synthesis of branched-chain amino
acids (Allison, 1969). This is accomplished by reductive carboxylation to produce the keto-acid homologue of the amino acid followed by introduction of the
amino group. Thus, isobutyric acid is reductively carboxylated to α-ketoiso-
valeric acid with subsequent amination to produce Val. Leu is produced from
2-methylbutyric acid and Ile from isovaleric acid by similar reactions. Some
populations use the branched-chain fatty acids as precursors for the synthesis of
branched long-chain fatty acids which are incorporated into cellular lipids
(Bryant, 1974).

Rumen protozoa do not have the biosynthetic capacity to synthesize amino
acids. The bacteria of the rumen are very important to the protozoa as sources
of nitrogen. Many of the protozoa, particularly the entodiniomorphs, ingest
bacteria, digest their proteins to amino acids, and resynthesize protozoan protein
from the amino acids (Coleman, 1975). Some of the protozoa are able to use
free amino acids as sources of nitrogen (Coleman and Laurie, 1977).

Although the ruminant does not require B vitamins in its diet, most of the major bacterial species in the rumen require a few B vitamins when grown in defined media. This indicates that some populations of rumen bacteria produce the B vitamins that others require. Two unusual growth factor requirements that apply only to bacterial interactions are requirements for heme and for coenzyme M. Heme is required for growth by some populations and is produced by others (Caldwell *et al.*, 1965; Reddy and Bryant, 1977). Requirements for heme for growth of human large-intestine anaerobes are also common (Sperry *et al.*, 1977). Coenzyme M appears to be unique to methanogens and to their metabolism. It is produced by some methanogens and used by others (Taylor *et al.*, 1974).

The nutritional requirements of rumen microorganisms are, therefore, related to the nutrients available to them in the more or less steady-state environment of the rumen. The composition of the food eaten by the animal gives very little indication of what individual species of rumen microorganisms require for growth. Many of the nutritional requirements are unique to the rumen or at least to it and similar habitats. In addition to the requirements mentioned previously, several rumen species have absolute requirements for sodium (Caldwell and Hudson, 1974), which is in high concentration in the rumen. Large amounts of sodium are present in ruminant saliva. Many populations in the rumen require carbon dioxide for growth (Dehority, 1971). Its prominence in the rumen has already been mentioned. It is of little use to cultivate rumen microorganisms on so-called rich media, e.g., blood, milk, or many of the complex commercial media that contain a variety of protein digests and yeast, liver, or soybean extracts. Although these media may contain some of the ingredients required by some of the major rumen populations, they are generally deficient in some of the important ingredients these organisms have available to them in the rumen. There is really nothing novel about the relationship between an organism's nutritional requirements and the composition of its natural habitat. It has long been a practice in microbiology to attempt to use habitat-simulating media to isolate organisms from a habitat, e.g., milk for lactic acid bacteria and blood for blood pathogens. The special feature of habitats that contain a complex microbial community is that the nutrient composition of the steady-state habitat greatly differs from the composition of the nutrients used to feed the system.

6. Fermentation Interactions

6.1. Food Chains

There are a few examples where products of the energy metabolism of one rumen microbial population serve as the sources of energy for other populations. The most prominent example is the chain that leads to the formation of methane. Rumen methanogenic bacteria use either hydrogen and carbon dioxide or formate as substrates for the production of methane (Hungate, 1967; Hungate *et al.*, 1970). The reduction of carbon dioxide to methane and the essentially equivalent production of methane from formate provide energy for the growth of the rumen methanogens. Methanogens from other environments produce methane from acetate or methanol (Mah *et al.*, 1977; Zeikus, 1977), but these are not substrates for ruman methanogenesis. There are no methanogens known that use carbohydrates, amino acids, purines or pyrimidines, or any substrates other than those cited above as sources of energy. Carbohydrate-fermenting bacteria in the rumen produce the hydrogen and carbon dioxide and formate that the rumen methanogens use for the production of methane. Hydrogen and carbon dioxide are by far the major precursors of methane in the rumen (Hungate *et al.*, 1970).

Another food chain that is of occasional importance is the production of lactic acid by some rumen bacteria and its use by others. Ordinarily, lactate is not an intermediate in the overall fermentation. It is produced and used when animals are fed considerable quantities of easily digestible, starch-containing diets. Starch-fermenting species such as *S. bovis* and *S. ruminantium* are capable of producing considerable amounts of lactate. Lactate-using species, such as *Megasphaera elsdenii* and *Veillonella alcalescens*, are able to use lactate as an energy source. There are also lactate-fermenting strains of *S. ruminantium* present in the ecosystem.

6.2. Products of Polymer Hydrolysis

It is reasonable to emphasize the ability of organisms to directly use nutrients that enter the ecosystem as sources of energy. Most of the rumen bacteria and protozoa can use one or another of the plant carbohydrates ingested by the animal (Table III). There are some carbohydrate-fermenting species and strains of species that do not attack cellulose, starch, or hemicellulose (Table III). Since these organisms are present in high concentrations in the environment, it is likely that they feed on soluble carbohydrates produced by starch- and cellulose-fermenting populations. This of course assumes that they are not using other organisms' fermentation products or unidentified energy sources. Cross-feeding at the level of the soluble products of polymer hydrolysis is a poorly understood facet of the interaction between populations in the rumen and in other complex microbial ecosystems. A few general points will be discussed here to indicate the possible significance of these interactions and some directions to be taken in future research to clarify their nature and significance.

An obvious, fundamentally important, but often glossed-over feature of the rumen and other complex microbial ecosystems is that most of the important carbon and nitrogen sources are introduced as insoluble polymers. Extracellular enzymes have to be used to digest these polymers. Products are formed that enter the cells of the organisms that digest the polymers, but the products are also available to other species that do not participate in the primary events of hydrolysis. Competition can take place for the permeating monomeric or oligomeric products. Competition is not necessarily restricted to the organisms that digest polymers and those that do not. Competition for products can occur between polymer-attacking species. Obviously, competition cannot operate to the detriment of populations that have the primary tasks of initiating the digestion of high-molecular-weight nutrients. The community would not be able to survive such extreme competition. Sufficient cross-feeding can occur, however, to maintain species that cannot ferment polymers per se and to influence the dynamics of the digestion process. Important considerations in this competition are the nature of the soluble products of hydrolysis and the permeability of organisms in the ecosystem to these products.

An example of this cross-feeding is provided by the demonstration of the growth of cocultures of *S. ruminantium* and *B. succinogenes* on cellulose. *B. succinogenes* is cellulolytic, and *S. ruminatium* is not. When the two species are grown together with cellulose as the sole energy source, the selenomonad grows as well as the cellulolytic species (Scheifinger and Wolin, 1973). The nature of the soluble products of cellulose hydrolysis preferred by each organism is not known. Both species grow well together when cocultured on cellobiose or Glc, which both can use independently.

Cross-feeding of products from nonpermeating substrates occurs with substances other than carbohydrates. Lipids are hydrolyzed in the rumen to yield glycerol, which is fermented by glycerol-using organisms. The relationships between products of protein hydrolysis and nitrogen requirements were discussed previously. Cross-feeding of products of protein hydrolysis is no doubt important in protein metabolism, but specific intermediates have not been identified. Further investigations are necessary to identify the permeating products of the breakdown of all substrates. It is the permeability of dominant species to specific products of hydrolysis that is important in many microbial ecosystems.

6.3. Decarboxylation of Succinate

Figure 3 depicts a scheme for the overall fermentation of plant polysaccharides by the rumen microbial community. Portions of the scheme have already been discussed. Cellulose and starch are hydrolyzed to permeable products. These are fermented by the hydrolyzing populations and populations that feed on the hydrolysis products. Fermentation produces the end products of the rumen fermentation, and major and minor extracellular intermediates. The minor intermediates include lactate and formate; lactate is only occasionally formed on specific diets, and formate does not appear to be formed in large

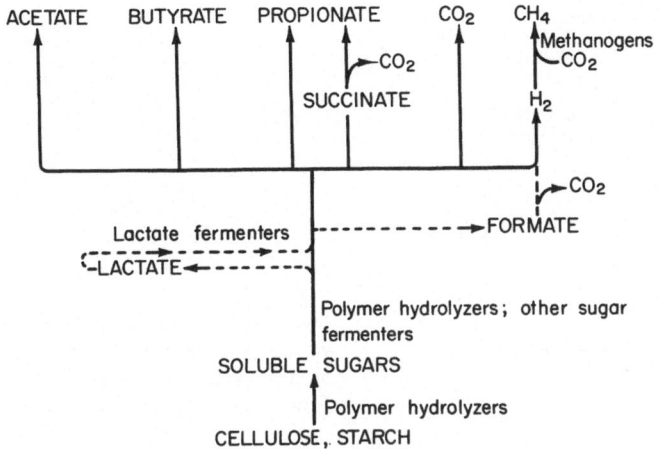

Figure 3. Intermediates and products of the rumen fermentation of plant polysaccharides; the dashed line represents minor pathways.

amounts. Lactate, when formed, is fermented to the same products formed from carbohydrates. Formate is converted to hydrogen and carbon dioxide, which are then converted to methane. Hydrogen and carbon dioxide from formate and from other, more important intracellular intermediates are important precursors of methane. Succinate is an important extracellular intermediate in the overall rumen fermentation. It is decarboxylated essentially as fast as it is formed (Blackburn and Hungate, 1963). The extracellular pool of succinate is extremely small.

Monocultures of many important bacterial species of the ecosystem produce succinate as a major product of carbohydrate fermentation. Very few species produce propionate. The major propionate producers in the bovine rumen are *S. ruminantium* and *M. elsdenii.* The former species forms propionate from carbohydrate by a pathway similar to that used by the propionibacteria, the randomizing pathway. Succinate is an intermediate (intracellular) of the randomizing pathway. It can be decarboxylated by a series of enzymes to propionate and carbon dioxide. *S. ruminantium* can decarboxylate succinate to propionate and carbon dioxide. Experiments with cocultures of *B. succinogenes* or *R. flavefaciens* and *S. ruminantium* demonstrate that the selenomonad is able to grow on carbohydrates produced from cellulose by either cellulolytic species. The selenomonad produces propionate from carbohydrate and decarboxylates the succinate produced by the other species to propionate. The net result is a coculture fermentation that produces propionate instead of succinate. The interaction is depicted in Fig. 4. The selenomonad obtains no apparent benefit from carrying out the decarboxylation of succinate produced by the succinate-

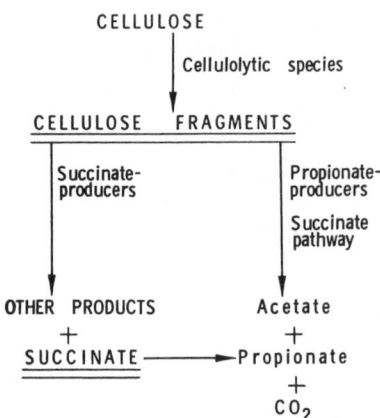

Figure 4. Interaction between species involved in the production of succinate and its decarboxylation to propionate in the rumen. From Wolin (1974); reprinted by permission.

producing species of the rumen. It is not clear why this two-species system has developed for the production of propionate in the rumen.

M. elsdenii does not use the randomizing pathway for production of propionate and presumably does not decarboxylate succinate to propionate. *V. alcalescens* uses the randomizing pathway and does decarboxylate succinate. Although found in the rumen of sheep, *Veillonella* strains have not been shown to be important in cattle. Because *Veillonella* strains do not use carbohydrates or glycerol but are able to ferment lactate, they are probably important in sheep only when the diet leads to conditions for the production of significant amounts of extracellular lactate.

Recent experiments suggest that the interactions necessary for propionate formation in the rumen may not occur in the human intestinal tract. Many of the dominant nonsporing anaerobes, particularly *Bacteroides*, produce succinate as a major product of carbohydrate fermentation in monocultures. When supplemented with vitamin B_{12}, however, several of these same species produce propionate instead of succinate, or at least less succinate and more propionate (Table VI). One of the important enzymes in the formation of propionate from succinate is methylmalonyl-coenzyme A isomerase, which requires a coenzyme form of vitamin B_{12} for activity. Human intestinal *Bacteroides* apparently can synthesize the apoenzyme but cannot synthesize the vitamin B_{12} necessary for synthesis of the coenzyme. Conversion of succinate to propionate in the large intestine is probably a single-organism, intracellular process that depends on the presence of vitamin B_{12}. The reason for the apparent difference between the methods of propionate formation in the rumen and the human large intestine is not clear. It is of some interest to note that propionate formation is absolutely essential to the ruminant because propionate is the only gluconeogenic volatile acid in the rumen; i.e., it is the only one the animal can use for the synthesis of

Table VI. Requirement for Vitamin B_{12} for Propionate Formation
by Human Intestinal *Bacteroides*[a]

Source and organism	+ Vitamin B_{12}[b]		– Vitamin B_{12}	
	Succinate	Propionate	Succinate	Propionate
	mol/100 mol hexose[c]			
Human intestine				
Bacteroides eggerthii	79	27	115	0
B. fragilis	33	49	92	9
B. splanchnicus	33	79	80	0
Rumen				
B. succinogenes	77	0	n.d.[d]	n.d.
B. ruminicola	130	0	n.d.	n.d.
Ruminococcus flavefaciens	68	0	n.d.	n.d.

[a] Miller and Wolin (1979); reprinted by permission.
[b] 20 ng/ml.
[c] Glucose except for *R. flavefaciens*, which received cellobiose.
[d] n.d., Not determined.

glucose. Glucose is as important in a ruminant's metabolism as it is in the metabolism of monogastric animals. There is no evidence that propionate production in the large intestine is of any direct significance to humans. Perhaps the two-organism system is more effective as a propionate-producing system and has evolved because of the essential role of propionate in ruminant metabolism.

6.4. Methanogenesis

Various aspects of methanogenesis have been reviewed recently, including microbiological, biochemical, and ecological perspectives (Mah *et al.*, 1977; Zeikus, 1977). A few comments will be added here to emphasize some differences between the rumen and other ecosystems and to emphasize the role of methanogenesis in maintaining low partial pressures of hydrogen in the rumen and other systems.

Two major pathways are used in nature for the formation of methane. One involves the reduction of carbon dioxide by hydrogen, and the other involves the decarboxylation of acetate to methane and bicarbonate. Methane is produced from the methyl group of acetate by the latter pathway, and carbon dioxide is not reduced. Not only is the carbon of methane derived from the methyl group of acetate, but the three hydrogens of the methyl group are transferred intact to the final product. The contributions of carbon dioxide reduction by hydrogen and acetate decarboxylation to methanogenesis in an ecosystem can be determined by measuring the amount of methane formed from radioactive carbon dioxide and the amount formed from radioactive, methyl-labeled acetate. Labeling studies in the rumen show that all the methane is produced by reduction of carbon dioxide. Decarboxylation of acetate does not occur. In waste decomposition and anaerobic sediments, about 70% of the methane produced is derived from the methyl group of acetate and 30% is from reduction of carbon dioxide (Mah *et al.*, 1977). The probable reason for the lack of production from acetate in the rumen is the slow growth of the bacteria that produce methane from acetate. Their generation times appear to be too long to permit their establishment in light of the dilution rate of the rumen system. In the other environments, however, turnover is much slower, and the acetate-using organisms are able to grow and produce methane.

Production and loss of methane is usually considered as a loss of energy for the ruminant. Approximately 10 to 15% of the energy of ingested food is lost by eructation of methane. It is possible to imagine a fermentation that would permit the accumulation of acetate, propionate, and carbon dioxide (like the glucose fermentation of the propionibacteria) that would be more beneficial to the animal. No energy would be lost as eructated gas, and the volatile acids are those the animal already uses. Any interference with the production of methane that leads to the accumulation of hydrogen, however, would not be beneficial because there would be an even greater loss of energy. The reduction of carbon dioxide by hydrogen does provide energy for the synthesis of the protein and other cellular constituents of the methanogens which are of benefit to the animal. Nevertheless, considerable attention has been given by animal nutritionists to the possibility of using inhibitors of methanogenesis with the hope of capturing the lost gaseous energy in nongaseous fermentation products the animal can use advantageously (Demeyer and Van Nevel, 1975).

The use of hydrogen by methanogens is extremely important in all methanogenic fermentations. The capacity of methanogens to use hydrogen is immense. Methane is produced by rumen contents at a rate of approximately 5.5 μmol/g per hour (Carroll and Hungate, 1955). Since 4 mol of hydrogen are used to form 1 mol of methane, the rate of hydrogen utilization is 4 times the rate of production of methane. The hydrogen-production rate is the same as the hydrogen-utilization rate. (The amount of methane produced and hydrogen used per day in a 500-kg cow is approximately 200 and 800 liters, respectively.) No hydrogen accumulation occurs in the rumen; the partial pressure of hydrogen is approximately 10^{-4} atm (Hungate, 1967). (The partial pressure of methane is about 0.35 atm.) Very low partial pressures of hydrogen are also found in other environments where methane is a major product of fermentation. The partial pressure of hydrogen has an important influence on the course of the rumen fermentation and other methanogenic fermentations.

6.5. Hydrogen Gas and Fermentation

Some bacteria are able to form hydrogen from reduced pyridine nucleotides (PNH), but the accumulation of hydrogen inhibits the reaction. The equilibrium constant strongly favors the reverse reaction, the production of PNH from oxidized pyridine nucleotides and hydrogen. Removal of hydrogen from the environment, however, permits the expression of hydrogen formation from PNH. Other major mechanisms for microbial formation of hydrogen include formation from formate and pyruvate. These routes are not susceptible to reversal by hydrogen. The importance of the hydrogen-inhibited and -uninhibited routes on the fermentation of glucose can be illustrated by the fermentation pathway of *R. albus* (Glass *et al.*, 1977) (Fig. 5). It ferments glucose by the Embden–Meyerhof–Parnas pathway to pyruvate and NADH. Acetyl-CoA, hydrogen, and carbon dioxide are formed from pyruvate in a ferredoxin-dependent reaction. The catalytic amounts of NADH produced in the cell must be reoxidized. They can be reoxidized by reduction of acetyl-CoA to acetaldehyde, which, in turn, is reduced to ethanol by NADH. The acetyl-CoA that is not reduced is converted to acetate. These reactions lead to a balanced fermentation of glucose to ethanol, acetate, hydrogen, and carbon dioxide. *R. albus* can also reoxidize NADH with the production of hydrogen with another ferredoxin-dependent system but only when hydrogen is removed from its surroundings. This happens in the rumen when the organism is growing in the presence of methanogens or in artificial systems when the organism is cocultured with a hydrogen-using species (Iannotti

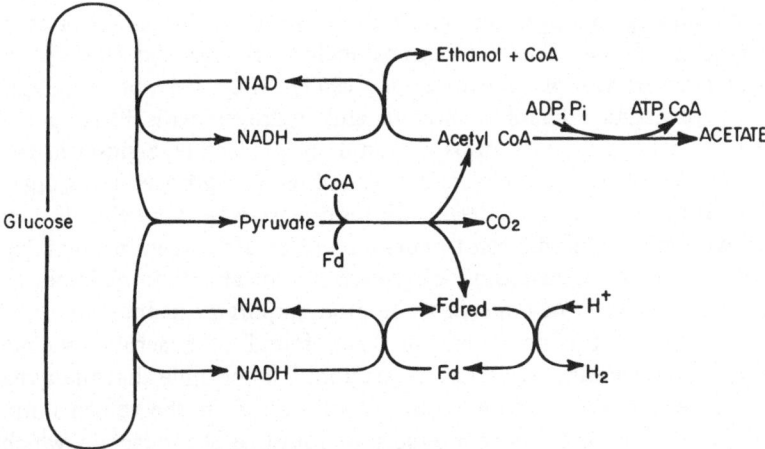

Figure 5. Pathway of fermentation of glucose by *Ruminococcus albus*. Fd, Ferredoxin. From Glass *et al.* (1977); reprinted by permission.

et al., 1973). If NADH is used to produce hydrogen, it cannot be used to produce ethanol. All of the acetyl-CoA formed from pyruvate becomes available for the production of acetate. When hydrogen is removed, the balanced fermentation products are acetate, carbon dioxide, and the removed hydrogen, and ethanol is no longer a product. Ethanol is not produced from carbohydrate in the rumen (Moomaw and Hungate, 1963), despite the fact that important species such as *R. albus* produce ethanol as a major fermentation product in pure cultures (where hydrogen is usually allowed to accumulate).

Similar effects of hydrogen utilization on the formation of succinate by *R. flavefaciens* (Latham and Wolin, 1977) and propionate and lactate by *S. ruminantium* (Chen and Wolin, 1977) have been demonstrated. The general principle is the same as for the *R. albus* fermentation. Succinate, propionate, and lactate are electron-sink products of the respective carbohydrate fermentations of *R. flavefaciens* and *S. ruminantium*, as is ethanol for *R. albus*. They can reoxidize NADH formed during glycolysis by forming the electron-sink products, succinate, lactate, propionate, or ethanol, from pyruvate. When the partial pressure of hydrogen is kept very low by its removal by a methanogen, production of any of the electron-sink products greatly diminishes, and the production of acetate from pyruvate increases.

More detailed discussions of these interactions are available in other publications (Wolin, 1974, 1975, 1976), and only a few highlights will be discussed here. The organisms mentioned above have alternative mechanisms for reoxidizing NADH, and their fermentation products will differ depending upon the partial pressure of hydrogen in the environment. Some organisms are known that do not use carbohydrates and do not have alternatives. They can only reoxidize NADH by forming hydrogen and can only do so when the partial pressure of hydrogen is kept low. These organisms include S organism (Reddy *et al.*, 1972a) and desulfovibrios (Bryant *et al.*, 1977). They can obtain energy for growth by oxidizing ethanol to acetate and hydrogen with PNH-dependent enzyme systems, and desulfovibrios can similarly use lactate. Significant growth occurs only when the organisms are cocultured with a hydrogen-using organism (a methanogen). Alternatively, desulfovibrios can grow by themselves if they are provided with sulfate in addition to ethanol and lactate. Organisms that depend on interactions with methanogens for growth do not appear to be important in the rumen ecosystem. They appear to be more important in the complete bioconversion systems for producing methane found in anaerobic sediments, swamps, and in anaerobic waste decomposition. It is possible that the dynamic balance between hydrogen formation and utilization in the rumen normally avoids the formation of pulses of products such as ethanol and lactate, which are energy sources for the growth of organisms like S organism and desulfovibrios. Closer juxtaposition between interacting organisms, better mixing, and the generally greater constancy of nutrients and the environment of the rumen may mitigate against the more complex food chains of the other ecosystems.

In addition to the interactions between S organism or desulfovibrios and methanogens, interactions such as those described for the rumen organisms probably occur in concert with the dependent interactions in other ecosystems. Carbohydrate-fermenting clostridia can form hydrogen from NADH (Jungermann *et al.*, 1973), and interactions between a thermophilic, cellulolytic *Clostridium* and a thermophilic methanogen (Weimer and Zeikus, 1977) are similar to those described for *R. albus*. Interaction between hydrogen formation and utilization is not necessarily restricted to hydrogen utilization by methanogens. The reduction of carbon dioxide to acetate is an important mechanism for using hydrogen in the ceca of rodents (Prins and Lankhorst, 1977), and sulfate reduction by hydrogen by sulfate-reducing bacteria (Badziong *et al.*, 1978) is another potentially important mechanism for the depletion of hydrogen in anaerobic ecosystems. Production of hydrogen from NADH is also not restricted to carbohydrate-, ethanol-, and lactate-using microorganisms. The products of Gly fermentation by *Diplococcus glycinophilus* are altered by the partial pressure of hydrogen in a manner that is consistent with an effect on alternative modes of oxidizing NADH (Cardon and Barker, 1947; Wolin, 1976). The use of formate and probably pyruvate by S organism involves reoxidation of NADH to hydrogen (Reddy *et al.*, 1972b; Wolin, 1976).

Additional research is necessary to obtain a clearer perspective of the role of hydrogen formation and utilization on the metabolism of individual microorganisms and on the flow of carbon in anaerobic ecosystems. It is clear, however, that these interactions can profoundly influence the types of products formed in the environment and, therefore, the actual substrates available for the establishment of specific food chains and the relevant microorganisms. For organisms that do not possess alternative routes of oxidation of NADH, interaction is essential for the production of energy and for their survival. Those organisms that do have alternatives may obtain more energy for themselves when they produce acetate rather than electron-sink fermentation products from pyruvate. For example, more ATP is produced from glucose when acetate rather than ethanol or lactate is the fermentation product. The increased ATP yield may be advantageous to an organism if the amount of its energy supply in its environment limits its growth.

6.6. Growth Rate and Fermentation

It is difficult to analyze the kinetics of growth of individual populations in the rumen ecosystem. Cognizance of the general subject of growth kinetics indicates, however, that there are many factors in a complex ecosystem that can control the rate of growth of microogranisms. These factors include the rates of hydrolysis of polymeric carbon and nitrogen sources, concentrations of the permeable macronutrients (e.g., carbon, energy, and nitrogen sources) actually used by the organisms, concentrations of any potentially growth-limiting micronutrients (e.g., vitamins and trace metals), inhibitory agents, and pH. Since protozoa depend to a significant extent on ingestion of intact bacteria for nutrients, rates of ingestion and intracellular digestion are important in determining the rates of growth of protozoa. Various aspects of the effects of growth rate, substrate concentration, and pH on fermentation have been reviewed (Hobson et al., 1974). One example of the influence of growth rate on fermentation is the demonstration that S. ruminantium produces mainly propionate, acetate, and carbon dioxide from glucose when grown at slow growth rates in a glucose-limited chemostat but switches to the production of increasing amounts of lactate as the growth rate increases (Hobson and Summers, 1967; Scheifinger et al., 1975). These pure culture experiments suggest that the fermentation of S. ruminantium in the ecosystem may vary depending on the rate it is supplied with a source of fermentable carbohydrate. This example is used to illustrate the point that many factors other than the partial pressure of hydrogen can alter the course of fermentation of individual species. Since it is difficult to directly measure growth rate and the environmental parameters that influence it in the ecosystem, studies of factors that regulate pure cultures can at least help define their potential for carrying out different kinds of fermentations and the conditions that might regulate the fermentations in the ecosystem.

6.7. The Overall Fermentation

The fermentations of individual populations and the already described interactions between populations produce acetate, propionate, butyrate, methane, and carbon dioxide from plant carbohydrates. The stoichiometry of the fermentation of glucose, the monosaccharide unit of cellulose and starch, by the community has been deduced by independent methods (Wolin, 1960; Hungate, 1966) and is approximately

$$57.5 \, (C_6H_{12}O_6) \longrightarrow 65 \text{ acetate} + 20 \text{ propionate} + 15 \text{ butyrate} + 60 \, CO_2$$
$$+ \, 35 \, CH_4 + 25 \, H_2O$$

The major pathway for hexose fermentation in the rumen is the Embden-Meyerhof–Parnas pathway, the major pathway of most of the individual species of carbohydrate-fermenting rumen microorganisms. This subject and the pathways of pyruvate metabolism and product formation of carbohydrate-fermenting rumen bacterial species have been reviewed recently (Miller and Wolin, 1979). As previously discussed, hydrogen utilization would tend to decrease the amount of propionate and eliminate the lactate and ethanol formed by isolated single species with corresponding increases in acetate in the ecosystem. There is no evidence that hydrogen utilization influences the formation of butyrate in the rumen or by pure cultures of one of the major butyrate-forming species, *B. fibrisolvens* (Wolin, unpublished data). It should be emphasized that the concentrations of individual species in the rumen may also be a major factor in determining the proportions of products formed from carbohydrate. For example, *B. succinogenes* produces succinate and acetate from cellulose but does not produce hydrogen. Conditions that favor *B. succinogenes* over the cellulolytic *R. albus* would favor the production of propionate (by decarboxylation of succinate) rather than production of methane and acetate (by *R. albus*; see Fig. 5). The relationship between fluctuations in the concentrations of individual species and the relative proportions of acid products has not been explored.

7. Summary and Conclusions

This review has focused on the rumen fermentation. In order to discuss the fermentation, it was necessary to provide an overview of the ruminant as a fermentation factory. This was possible because of the vast amount of available information about ruminant physiology, metabolism, and nutrition. Analogous information about other intestinal-tract systems is scarce, and relationships between microbial activities in the intestinal tract and host physiology, metabolism, and nutrition are not always clear. It was also possible to discuss the *in vivo* rumen microbial community and its activities, gross composition, environment, and dynamics because of the studies of many investigators. Again, there is a scarcity of similar studies of other intestinal-tract and other anaerobic ecosystems. Finally, it was possible to attempt to dissect the activities of the rumen community into activities of individual populations of bacteria and protozoa because many of them have been isolated and well characterized. Again, this is not always true for other intestinal-tract or other complex anaerobic ecosystems.

The rumen fermentation is of special importance to man. Domestic ruminants are factories for the conversion of solar energy, through the medium of plants, into very high-quality food, fiber, and leather. Ruminants can thrive on plants that are otherwise difficult to harvest and to process into something useful for man. Microorganisms in the rumen convert the cellulose and other carbohydrates, the nitrogen constituents, and trace elements of plants to short-chain volatile fatty acids and microbial cells. The animal uses the microbial products as sources of carbon, energy, protein, and vitamins. The ruminant and its microbial community interact to provide the proper environment for the necessary conversions. The microbial populations sustain each other by providing each other with essential nutrients.

The rumen fermentation itself is the result of interactions between many different microbial species. These interactions are not always apparent from conventional pure-culture studies—e.g., the ability of noncellulolytic organisms to use soluble oligosaccharides formed from cellulose by cellulolytic species, the decarboxylation of succinate produced by one species by another species, and the changes in fermentation products caused by the use of hydrogen by methanogenic bacteria.

The rumen system warrants further study not only because of its practical importance but also because it is an excellent system for studying phenomena that are of importance in other microbial ecosystems. Because of the existing data base and the remarkable constancy of the environment, the rumen microbial community is more amenable than most to studies of microbial interactions in complex ecosystems. We should, however, increase our efforts to analyze other systems using the strategies that have been successful for studies of the rumen. As with all complex ecosystems, it is clear that it is necessary to understand how the whole system operates in order to understand the contribution of its parts and that it is necessary to understand the contribution of the parts in order to understand the whole. Other systems may not be as well controlled as the rumen system and, therefore, less easily studied. It would probably be useful, however, to recognize that integration of studies of the community as a whole with studies of the populations of the community is not only a necessary but also a synergistic way of obtaining an understanding of the activities of an ecosystem.

ACKNOWLEDGMENT

The author's research is supported by research grant number AI-12461, awarded by the National Institute of Allergy and Infectious Diseases, U.S. Public Health Service, Department of Health, Education, and Welfare.

References

Allison, M. J., 1969, Biosynthesis of amino acids by ruminal microorganisms, *J. Anim. Sci.* 29:797–807.

Badziong, W., Thauer, R. K., and Zeikus, J. G., 1978, Isolation and characterization of *Desulfovibrio* growing on hydrogen plus sulfate as the sole energy sources, *Arch. Microbiol.* 116:41–49.

Bauchop, T., 1971, Stomach microbiology of primates, *Annu. Rev. Microbiol.* 25:429–436.

Becker, E. R., 1932, The present status of problems relating to the ciliates of ruminants and equidae, *Q. Rev. Biol.* 7:282–297.

Blackburn, T. H., and Hungate, R. E., 1963, Succinic acid turnover and propionate production in the bovine rumen, *Appl. Microbiol.* 11:132–135.

Bryant, M. P., 1959, Bacterial species of the rumen, *Bacteriol. Rev.* 23:125–153.

Bryant, M. P., 1974, Nutritional features and ecology of predominant anaerobic bacteria of the intestinal tract, *Am. J. Clin. Nutr.* 27:1313–1319.

Bryant, M. P., 1977, Microbiology of the rumen, in: *Duke's Physiology of Domestic Animals*, 9th Revised Edition (M. J. Swenson, ed.), pp. 287–304, Cornell University Press, Ithaca, New York.

Bryant, M. P., and Robinson, I. M., 1962, Some nutritional characteristics of predominant culturable rumen bacteria, *J. Bacteriol.* 84:605–614.

Bryant, M. P., Campbell, L. L., Reddy, C. A., and Crabill, M. R., 1977, Growth of Desulfovibrio in lactate or ethanol media low in sulfate in association with H_2-utilizing methanogenic bacteria, *Appl. Environ. Microbiol.* 33:1162–1169.

Caldwell, D. R. and Hudson, R. F., 1974, Sodium, an obligate requirement for predominant rumen bacteria, *Appl. Microbiol.* 27:549–552.

Caldwell, D. R., White, D. C., Bryant, M. P., and Doetsch, R. N., 1965, Specificity of the heme requirement for growth of *Bacteroides ruminicola, J. Bacteriol.* 90:1645–1654.

Cardon, B. P., and Barker, H. A., 1947, Amino acid fermentations by *Clostridium propionicum* and *Diplococcus glycinophilus, Arch. Biochem.* 12:165–180.

Carroll, E. J., and Hungate, R. E., 1955, Formate dissimilation and methane production in bovine rumen contents, *Arch. Biochem. Biophys.* 56:525–536.

Chen, M., and Wolin, M. J., 1977, Influence of CH_4 production by *Methanobacterium ruminantium* on the fermentation of glucose and lactate by *Selenomonas ruminantium, Appl. Environ. Microbiol.* 34:756–759.

Coleman, G. S., 1975, The interrelationship between rumen ciliate protozoa and bacteria, in: *Digestion and Metabolism in the Ruminant* (I. W. McDonald and A. C. I. Warner, eds.), pp. 149–164, University of New England Publishing Unit, Armidale, Australia.

Coleman, G. S., and Laurie, J. I., 1977, The metabolism of starch, glucose, amino acids, purines, pyrimidines and bacteria by the rumen ciliate, *Polyplastron multivesiculatum, J. Gen. Microbiol.* 98:29–37.

Dehority, B. A., 1971, Carbon dioxide requirement of various species of rumen bacteria, *J. Bacteriol.* 105:70–76.

Demeyer, E. I., and Van Nevel, C. J., 1975, Methanogenesis, an integrated part of carbohydrate fermentation, and its control, in: *Digestion and Metabolism in the Ruminant* (I. W. McDonald and A. C. I. Warner, eds.) pp. 366–382, University of New England Publishing Unit, Armidale, Australia

Eadie, J. M., 1962, The development of rumen microbial populations in lambs and calves under various conditions of management, *J. Gen. Microbiol.* 29:563–578.

el-Shazly, K., 1952, Degradation of protein in the rumen of the sheep. II. The action of rumen microorganisms on amino acids, *Biochem, J.* 51:647–653.

Glass, T. L., Bryant, M. P., and Wolin, M. J., 1977, Partial purification of ferredoxin from *Ruminococcus albus* and its role in pyruvate metabolism and reduction of nicotinamide adenine dinucleotide by H_2, *J. Bacteriol.* 131:463–472.

Hobson, P. N., and Summers, R., 1967, The continuous culture of anaerobic bacteria, *J. Gen. Microbiol.* 47:53–65.

Hobson, P. N., Bousfield, S., and Summers, R., 1974, Anaerobic digestion of organic matter, *CRC Crit. Rev. Environ. Control* 4:131–191.

Hungate, R. E., 1966, *The Rumen and Its Microbes*, 533 pp., Academic Press, New York.

Hungate, R. E., 1967, Hydrogen as an intermediate in the rumen fermentation, *Arch. Mikrobiol.* 59:158–165.

Hungate, R. E., 1975, The rumen microbial ecosystem, *Annu. Rev. Ecol. Syst.* 6:39–66.

Hungate, R. E., Bryant, M. P., and Mah, R. A., 1964, The rumen bacteria and protozoa, *Annu. Rev. Microbiol.* 18:131–166.

Hungate, R. E., Smith, W., Bauchop, T., Yu, I., and Rabinowitz, J. C., 1970, Formate as an intermediate in the bovine rumen fermentation, *J. Bacteriol.* **102**:389–397.

Hungate, R. E., Reichl, J., and Prins, R., 1971, Parameters of rumen fementation in a continuously fed sheep: Evidence of a microbial rumination pool, *Appl. Microbiol.* **22**:1104–1113.

Iannotti, E. L., Kafkewitz, D., Wolin, M. J., and Bryant, M. P., 1973, Glucose fermentation products of *Ruminococcus albus* grown in continuous culture with *Vibrio succinogenes*: Changes caused by interspecies transfer of H_2, *J. Bacteriol.* **114**:1231–1240.

Janis, C., 1976, The evolutionary strategy of the equidae and the origins of rumen and cecal digestion, *Evolution* **30**:757–774.

Jungermann, K., Thauer, R. K., Leimenstoll, G., and Decker, K., 1973, Function of reduced pyridine nucleotide-ferredoxin oxido-reductases in saccharolytic clostridia, *Biochim. Biophys. Acta* **305**:268–280.

Latham, M. J., and Wolin, M. J., 1977, Fermentation of cellulose by *Ruminococcus flavefaciens* in the presence and absence of *Methanobacterium ruminantium*, *Appl. Environ. Microbiol.* **34**:297–301.

Mah, R. A., Ward, D. M., Baresi, L., and Glass, T. L., 1977, Biogenesis of methane, *Annu. Rev. Microbiol.* **31**:309–341.

Miller, T. L., and Wolin, M. J., 1979, Fermentations of saccharolytic intestinal bacteria, *Am. J. Clin. Nutr.* **32**:164–172.

Moomaw, C. R., and Hungate, R. E., 1963, Ethanol conversion in the bovine rumen, *J. Bacteriol.* **85**:721–722.

Pittman, K. A., Lakshmanan, S., and Bryant, M. P., 1967, Oligopeptide uptake by *Bacteroides ruminicola*, *J. Bacteriol.* **93**:1499–1508.

Prins, R. A., and Lankhorst, A., 1977, Synthesis of acetate from CO_2 in the cecum of some rodents, *FEMS Microbiol. Lett.* **1**:255–258.

Reddy, C. A., and Bryant, M. P., 1977, Deoxyribonucleic acid base composition of certain species of the genus *Bacteroides*, *Can. J. Microbiol.* **23**:1252–1256.

Reddy, C. A., Bryant, M. P., and Wolin, M. J., 1972a, Characteristics of S organism isolated from *Methanobacillus omelianskii*, *J. Bacteriol.* **109**:539–545.

Reddy, C. A., Bryant, M. P., and Wolin, M. J., 1972b, Ferredoxin and nicotinamide adenine dinucleotide-dependent H_2 production from ethanol and formate in extracts of S organism isolated from "*Methanobacillus omelianskii*," *J. Bacteriol.* **110**:126–132.

Reichl, J. R., and Baldwin, R. L., 1975, Rumen modeling: Rumen input–output balance models, *J. Dairy Sci.* **58**:879–890.

Scheifinger, C. C., and Wolin, M. J., 1973, Propionate formation from cellulose and soluble sugars by combined cultures of *Bacteroides succinogenes* and *Selenomonas ruminantium*, *Appl. Microbiol.* **26**:789–795.

Scheifinger, C. C., Latham, M. J., and Wolin, M. J., 1975, Relationship of lactate dehydrogenase specificity and growth rate to lactate metabolism by *Selenomonas ruminantium*, *Appl. Microbiol.* **30**:916–921.

Scott, T. W., Ward, P. F. V., and Dawson, R. M. C., 1964, The fermentation and metabolism of phenyl-substituted fatty acids in the ruminant, *Biochem. J.* **90**:12–24.

Smith, R. H., 1975, Nitrogen metabolism in the rumen and the composition and nutritive value of nitrogen compounds entering the duodenum, in: *Digestion and Metabolism in the Ruminant* (I. W. McDonald and A. C. I. Warner, eds.), pp. 399–415, University of New England Publishing Unit, Armidale, Australia.

Sperry, J. F., Appleman, M. D., and Wilkins, T. D., 1977, Requirement of heme for growth of *Bacteroides fragilis*, *Appl. Environ. Microbiol.* **34**:386–390.

Taylor, C. D., McBride, B. C., Wolfe, R. S., and Bryant, M. P., 1974, Coenzyme M, essential for growth of a rumen strain of *Methanobacterium ruminantium*, *J. Bacteriol.* 120:974–975.

Weimer, P. J., and Zeikus, J. G., 1977, Fermentation of cellulose and cellobiose by *Clostridium thermocellum* in the absence and presence of *Methanobacterium thermoautotrophicum*, *Appl. Environ. Microbiol.* 33:289–297.

Wolin, M. J., 1960, A theoretical rumen fementation balance, *J. Dairy Sci.* 43:1452–1459.

Wolin, M. J., 1974, Metabolic interactions among intestinal microorganisms, *Am. J. Clin. Nutr.* 27:1320–1328.

Wolin, M. J., 1975, Interactions between the bacterial species of the rumen, in: *Digestion and Metabolism in the Ruminant* (I. W. McDonald and A. C. I. Warner, eds.), pp. 134–148, University of New England Publishing Unit, Armidale, Australia.

Wolin, M. J., 1976, Interactions between H_2-producing and methane-producing species, in: *Symposium on Microbial Production and Utilization of Gases* (H. G. Schlegei, G. Gottschalk, and N. Pfennig, eds.), pp. 141–150, E. Goltze KG, Göttingen, West Germany.

Wright, D. E., 1967, Metabolism of peptides by rumen microorganisms, *Appl. Microbiol.* 15:547–550.

Wright, D. E., and Hungate, R. E., 1967, Amino acid concentrations in rumen fluid, *Appl. Microbiol.* 15:148–151.

Zeikus, J. G., 1977, The biology of methanogenic bacteria, *Bacteriol. Rev.* 41:514–541.

3

Food as a Bacterial Habitat

A. HURST AND D. L. COLLINS-THOMPSON

1. Introduction

Food spoilage and food preservation have been a concern of man since the earliest beginnings of civilization. Some of the most important foods such as fruits, vegetables, meat, and milk are highly perishable, and their preservation challenged man's ingenuity. The essential need for food was the driving force from the earliest times in finding empirical ways of food preservation.

Even today, the cost of food spoilage must be enormous, but only the most vague estimates exist. These estimates generally deal with all forms of spoilage: bacterial, mold, vermin infestation, etc. J. H. Hulse of the Canadian International Development Research Centre thought that about one-quarter of food produced is lost.

The souring of milk may be regarded as spoilage or preservation, depending on one's point of view. Historically, it undoubtedly represents a preservation reaction. Under primitive conditions, it hardly ever fails; the resulting product is palatable, nutritious, and much longer lasting than the original article. Possibly, it is also safer; there is some evidence that the tubercle organism, which may be excreted in milk, may be destroyed by souring (Mattick and Hirsch, 1944).

Originally, many foods must have been discovered accidentally and then further developed and applied because they extended the shelf life of prime products, e.g., milk → sour milk → cheese. Fermented products of good organoleptic and keeping quality were discovered for meats, fish, and vegetables. The fermentation of these products relied on the indigenous flora harvested with the

A. HURST and D. L. COLLINS-THOMPSON • Bureau of Microbial Hazards, Health Protection Branch, Ottawa, Canada K1A 0L2.

prime product. Additionally, since at least early Egyptian times, man has used various preserving agents or determinants such as sodium chloride and nitrates. These helped to select bacterial populations important for flavor and color.

European civilization expanded rapidly into the rest of the world during the 19th century. This created social pressures for the availability of fermented foods with which the settlers and colonizers were familiar. Although it may be said that a beginning of the role of microbes in food spoilage and preservation can be dated to Pasteur's wine investigations in 1854 (and the use of heating as a means of removing undesirable organisms), early food technology, nevertheless, remained empirical. This was because the food industry operated in relatively small units. The flavor and texture of these products was acceptable largely because of the presence of the correct "natural" flora, and the level of the health hazard was acceptable also because at any one time relatively few persons were at risk. Cheddar cheese is a good example. It originated in a small English village but was manufactured in cheese factories in New York State in about 1870 (Kosikowski, 1977), to be followed later by manufacturing in Canada, New Zealand, and Australia. The lack of understanding of the processes involved at about the turn of the century is illustrated by the following practice told to one of the authors (A. H.) in the 1940s by a then-retired "dairy chemist": The lids of milk churns in southern England were fitted with holes to let out the "cowy" smell.

The traditional manufacturing systems worked well enough so long as the various intrinsic or extrinsic factors in the food did not vary. However, quite often intrinsic factors such as redox potential or antimicrobial constituents or extrinsic factors such as storage temperature changed, enabling invading organisms to enter the food microecosystem with resulting spoilage or risk of food poisoning. Such events are acceptable only if they affect small numbers of people. Increasing urbanization and growth of cities led to centralization of food production. With centralization, the potential for economic loss and mass food poisoning increased, and the "natural method" became less and less acceptable. Consequently, in the last 50 years, many fundamental studies concerning food microbiology have been carried out and applied to industrial processes. As a result, the flora associated with some products has changed, perhaps the most obvious being that which occurred in milk. In Western countries, city milk is always pasteurized and handled under refrigeration. The flora has changed from a mesophilic souring population to a psychrophilic–putrefactive one. Psychrophilic spoilage is much slower than souring, so that there has been a noticeable increase in shelf life. Also, pathogens from animal sources have been largely eliminated.

With the emphasis on safety and keeping quality during periods of the 1950s and 1960s, there was a serious possibility that the noble strong-tasting foods such as Hungarian salami or French Gruyère would become things of the past. Fortunately, because food is plentiful in the Western world, competition has forced manufacturers to pay attention to taste. This is being achieved by continual improvement and use of starter cultures in foods first rendered virtually sterile by sanitization or pasteurization processes. The most recent consumer preference is to buy a certain kind of cheese or sausage not only because it is wholesome and cheap but also because it tastes good. There is now a good chance that the advantage of large-scale manufacture will be combined with traditional quality.

The terms used in food ecology are the same as those employed in other biological studies and will not be defined in this article. An exception may be water activity (a_w) (Troller and Christian, 1978). This is the ratio of the water-vapor pressure of the food (p) relative to that of pure water (p_0)

$$a_w = p/p_0$$

Since equilibrium relative humidity (ERH) measures the same thing but is expressed as a percentage, ERH (%) = $a_w \times 100$, the water-vapor pressure of a food depends not on the total water but on the amount of free water. Free water is available for bacterial growth and is one of the principal factors which controls keeping quality.

Model systems for the study of food ecology exist but have been little used. A packet of food is a closed ecological system; the nutrients originally there are not renewed, and the waste products of metabolism accumulate. This can be modeled by batch cultures. In its wider sense, food ecology includes, for example, composting, the fate of microbes in decaying carcasses, etc; these are examples of open systems in which metabolic by-products are washed out. Continuous culture systems could be used as a model to study these ecological situations.

However, as food microbiology is an applied science, most of the literature has emphasized preservation and safety. Only scattered references exist dealing with mechanisms of population changes in food products. Many of these studies were done outside the food environment and may only represent a portion of the interactions between microorganisms in foods.

We will attempt in this review to classify foods according to ecological considerations, briefly discuss those microorganisms which have food significance, and give specific examples of microbial associations and successions in foods. Some important foods, such as sauerkraut and oriental fermentations, are only mentioned in passing. This was done to keep the chapter within reasonable proportions. Also, we made no attempt to cover the literature fully but, as far as possible, referred to review articles and books from which the interested reader may obtain further references.

2. The Major Bacterial Groups

2.1. Enterobacteria

The enterobacteria are gram-negative rods, aerobic and facultatively anaerobic, catalase positive, and oxidase negative; they may be motile and produce acid and/or gas from glucose and other carbohydrates. Many members of this group can also reduce nitrates to nitrites (Buchanan and Gibbons, 1974). The genera important in food are *Escherichia*, *Citrobacter*, *Salmonella*, *Shigella*, *Klebsiella*, *Enterobacter*, *Serratia*, *Proteus*, and *Yersinia*. These organisms are widely distributed in nature, their chief habitat being plants and animals. Several members of this family are animal pathogens.

Because many of these genera are found in the gut and fecal material of warm-blooded animals, their presence in the microflora of foods is important. They are most conveniently determined as coliforms or fecal coliforms. The coliform group is defined as those members of this family which are capable of fermenting lactose, with the production of acid and gas, at 37°C. The fecal types are further classified by use of incubation temperatures of about 45°C and biochemical tests such as indole production. Organisms included in the coliform test are *Escherichia*, *Citrobacter*, *Klebsiella*, and *Enterobacter*. The coliform test, originally developed from its use in water microbiology, is used extensively in food microbiology as an indication of sanitation. The presence of a large number of these organisms in foods is often taken to indicate fecal contamination. They also warn of the possible presence of pathogens such as *Salmonella* and *Shigella*. A belief exists that the higher the coliform count, the greater the possibility of pathogens in the food. In most cases, however, this relationship has never been proven. Cases are known where foods contained salmonellae but few or no coliforms. The reciprocal is also true: There are cases where high levels of fecal coliforms have been found (e.g., cheese) but no salmonellae! In cheese, low levels of fecal coliforms present in the milk can develop rapidly if the starter culture is slow in lowering the pH. Coliforms can reach 10^4/g of curd, and this is not related to fecal contamination. The coliform test has some benefits, especially since the discovery of enteropathogenic *Escherichia coli*, but it should be used only as a sanitation index.

Frazier (1968) has listed some other characteristics of the enterobacteria which are considered important in food microbiology. Included in this list are their ability to grow over a wide temperature range, the ease with which they can use a range of carbohydrates as energy sources, and their ability to use simple organic compounds as nitrogen sources. Their ability to compete successfully with other microorganisms is limited, especially in foods which contain 3–4% sodium chloride or where the pH is below 5.5.

Several members of this family are capable of growth at temperatures below 5°C. Psychrotrophic coliform bacteria, including *Klebsiella*, *Enterobacter*, and *Citrobacter*, grow in raw or pasteurized milk at refrigeration temperatures. Of these, *Klebsiella* grows especially well (Panes and Thomas, 1968), and *Klebsiella pneumoniae* is often present in refrigerated dairy products (Schiemann, 1976). The growth of coliforms in a number of products, including meats and vegetables stored at refrigerated temperatures, has been reported (Eddy and Kitchell, 1959). The growth rates, however, are lower than those of the pseudomonads, and the coliform bacteria never become the dominant group, accounting for 5-20% of the population.

Yersinia enterocolitica, an organism associated with human disease, is capable of growth below 5°C (Lee, 1977). It has been isolated from vacuum-packaged meats in sufficient numbers to suggest that it is capable of competing well with the other organisms at refrigeration temperatures.

Studies by Goepfert and Kim (1975) of anaerobically and aerobically packaged ground beef showed that at temperatures of less than 7°C, *E. coli* and *Salmonella* were poor competitors and showed little growth within the mixed flora. At temperatures of more than 10°C, however, these organisms grew and competed successfully. In comminuted meat products, like the British fresh sausage, the temperature is not the sole factor limiting growth. Studies by Dowdell and Board (1968) showed that coliforms in the sausage were not able to grow, even at 20°C. In this situation, ingredients such as sodium chloride may well have been inhibitory, although the antagonism between the coliforms and the natural flora of the sausage cannot be ruled out.

Production of bacteriocins is a property of several families of bacteria, including those of the enterobacteria. Colicins are produced by individual strains of *E. coli*, *Salmonella*, and *Shigella* and are bactericidal in nature (Ivanovics, 1962). The inhibitory action of colicins is restricted to a limited number of related species. Colicins are proteins that adsorb to the cellular surface of *E. coli* or related species and cause breakdown of macromolecules such as DNA. Synthesis and release of colicin kills the producer organism. However, transfer of this property occurs within the members of the same family.

The role of colicins in food microbiology may be academic. It is almost certain that strains of colicinogenic *Escherichia* or *Salmonella* are found in the human intestinal flora and therefore must find their way into the food chain. What effect this has on the control of specific microflora in foods is unknown. Tadd and Hurst (1961) fed colicinogenic *E. coli* to pigs in an effort to control intestinal disease caused by other strains of *E. coli*. Although the colicinogenic strain was rapidly established in the pigs' intestines, this did not suppress the pathogenic types and did not influence the frequency of occurrence of the disease.

Various other forms of antagonism between the coliforms and enteric pathogens have been reported (cited by Thimann, 1963). Examples of amensalism with the coliform group have been cited by Alexander (1971). When *E. coli*, *Klebsiella*, and *Shigella* were grown in mixed culture, only the *E. coli* and *Klebsiella* were able to reach cell densities similar to those reached by these organisms in pure culture. Antagonism between *Salmonella* and pseudomonads isolated from poultry has been reported by Oblinger and Kraft (1970). Such inhibition was related to pigment production by the pseudomonads. Inhibition of *Staphylococcus aureus* by coliform bacteria is a well-known phenomenon (Hurst, 1973). Many of these studies suggest inhibition via antibiotic production or competition for nutrients.

2.2. Pseudomonads

The genera *Pseudomonas*, *Acinetobacter* (which includes organisms originally classified as *Achromobacter*), *Alcaligenes*, *Moraxella*, and *Flavobacterium* are known to play an important role in the breakdown of food substrates at low temperatures. These organisms are found on the surface of fresh meats, poultry, fish, eggs, vegetables, and fruits. The species most frequently associated with food are *Pseudomonas fragi*, *P. fluorescens*, *P. putrefaciens*, *P. aeruginosa*, *P. tralucida*, and *P. ovalis*. Factors that make this group important are listed in Table I.

To maintain a longer shelf life, many perishable foods are stored at temperatures of 0-5°C. This temperature range permits growth of psychrophilic pseudomonads. The phenomenon of low-temperature growth is still not understood. Numerous studies have examined pigment production, enzyme activities, and cellular properties of the pseudomonads at low temperatures (Witter, 1961). Some *Pseudomonas* species produce pigments and lipolytic and proteolytic enzymes preferentially at temperatures below 15°C. Enzyme-kinetic studies have suggested that rate-limiting enzymes or enzyme systems may be involved, but such evidence has proven inconclusive. Psychrophilic pseudomonads tend to have a shorter lag phase than the mesophilic organism in the temperature ranges

Table I. Factors That Make *Pseudomonas* spp.
Important in the Breakdown of Foods

1. Growth at low temperatures (−10° to +5°C)
2. Production of lipolytic enzymes
3. Production of proteolytic enzymes
4. Ability to utilize simple nitrogenous foods
5. Ability to produce a variety of products which affect flavor
6. Aerobic metabolism enabling rapid surface growth

where both are capable of initiating growth. In consequence, at 0-5°C the pseudomonads are capable of outcompeting any of the mesophilic organisms present in foods. Two interesting concepts have been developed to compare the behavior of psychrophilic and mesophilic organisms in foods. These are the temperature coefficient (Q_{10}) and the mathematically related characteristic (μ). Study of these concepts has attempted, unsuccessfully, to relate psychrophily to low-temperature growth and enzyme activity. So far, chemical activity alone has not been able to fully explain low-temperature growth (Elliot and Michner, 1965).

The effect of temperature and growth rates on lipid structure was reviewed by Rose (1962). It was observed that some pseudomonads capable of growing at low temperatures contained a higher proportion of unsaturated fatty acids in the lipid portions of the cell than did mesophiles. Recent studies comparing the unsaturated fatty acid composition of *P. fluorescens* and the mesophilic *E. coli* grown at different temperatures indicated that the phenomenon of low-temperature growth cannot be accounted for in terms of the unsaturation of the fatty acids (Gill and Suisted, 1978).

Classification of the pseudomonads recognized their ability to utilize non-carbohydrate sources for energy. This important factor makes them competitive when they grow on low-carbohydrate substrates such as meat and fish. Studies by Shelef (1977) with beef and added glucose (2-10%) showed that the growth of predominant nonpigmented pseudomonads was hindered in the presence of added carbohydrate. Under normal growth conditions on a high-protein medium, the metabolic activities of the pseudomonad during the logarithmic phase tend to increase the surface pH toward the alkaline side. Addition of glucose to the ecosystem counteracts this, and the pH decreases to about 5.5.

Shelef (1977) also suggested that the addition of glucose produced a change in dominance from the pseudomonads to lactic acid bacteria (LAB). The pH balance appears to be critical and below pH 5.0, the LAB succeed. Control of the pseudomonads on a meat substrate has been achieved by introducing high levels of LAB into the ecosystem (Reddy *et al.*, 1970). In the presence of such starters, inhibition of the pseudomonads takes place due to a rapid pH drop and the presence of other metabolites such as peroxides (Gilliland and Speck, 1975).

Correlation of the oxidation-reduction potential (E_h) and the growth of the pseudomonads is poorly documented in the literature. It appears that definite E_h values can determine if a given organism can invade, survive, and grow in a given ecosystem (Oblinger and Kraft, 1973). Studies with *P. fluorescens* showed that at maximum growth rate the E_h reached its lowest point. This point becomes unfavorable for the aerobic organisms, and it is at this time that competition can arise or a succession process is set up with other facultative or anaerobic organisms. The studies by Oblinger and Kraft (1973) using cultures of *Salmonella* and *P. fluorescens* support this point. The growth of the pseudo-monads gradually lowers the E_h, maintaining their competitive ability. On the other hand, the less aerobic salmonellae establish favorable conditions in the ecosystem by rapidly reducing the E_h. In mixed cultures, at low E_h values, salmonellae appear to gain some measure of competitive edge against the pseudomonads. Further work using food substrates is needed to understand the relationship between E_h, competition, and succession relationships in foods.

Several studies have indicated that the pseudomonads have extracellular lipases (Witter, 1961; Bala *et al.*, 1977). Many of the lipases are preferentially produced at low temperatures (less than 15°C) (Alford *et al.*, 1964). The break-down of lipids in substrates such as red meats has been studied more from the point of view of flavor and color changes rather than ecological changes. Little work has been done to determine to what extent organisms benefit from the by-products of lipase activity. Some lipases, such as that from *P. fragi*, are fairly specific in their action, attacking primarily the 1 position of the triglyceride, although some action occurs in all three positions. Some specificity also exists as to the type of triglyceride attacked. In terms of supplying energy needs, the breakdown of fats does provide intermediates which can be utilized. Several reports also exist, including that of Kato and Shibasaki (1975), that certain fatty acids have antimicrobial activity. Such activity appears to be more pro-nounced against gram-positive bacteria, with little or no effect against *P. aeruginosa*. Fatty acids such as caproic and lauric and monoglycerides (monocap-rin and monolaurin) were active against *Micrococcus lysodeikticus* and *S. aureus*. Both microbes have been found in low numbers in raw meat and poultry.

The significance of proteolytic enzymes of the pseudomonads is well known for fish, milk, and meat. Studies by Shewan and Jones (1957) with fish showed that a steady increase in certain amino acids occurred during storage at 0°C. Lysine appeared to be generated at a higher rate than other amino acids such as glycine. Pseudomonads also produce extracellular lytic factors. Whiteside and Corpe (1969) isolated a proteinase that was active against *Chromobacterium violaceum*, another gram-negative bacterium. This lytic factor has also been shown to be active against *S. aureus* and *Sarcina lutea* (Zyskind *et al.*, 1965; Collins-Thompson *et al.*, 1973). This lytic factor contains lipases and mucopetide-*N*-acetylmuramic hydrolase. Such enzyme activities are not, however, limited to the pseudomonad group (Ghuysen, 1968).

Another property of the pseudomonads is the production of trimethylamine (TMA) from trimethylamine oxide (TMAO). Studies by Watson (1939) sug-gested that TMA produced in fish is a result of energy-yielding reactions between TMAO and appropriate hydrogen donors, especially with intermediates such as pyruvate, in an O_2-limiting ecosystem. Typical of the reactions involved would be pyruvate to lactate. The production rate of TMA reaches its maximum during the logarithmic phase of growth (i.e., at a high O_2-demand state).

Several investigations concern the public health significance of pseudomonads. Inhibition of *S. aureus* has been reported without the nature of the inhibition being elucidated (Defigueiredo and Splittstoesser, 1976). Many of the nonspecific reactions include outcompeting for nutrients and production of antibiotic substances. Unfortunately, several of these studies were done with broth cultures and not with a food substrate. Studies carried out by Miller *et al.* (1973) on fish revealed a host of potentially inhibitory intermediates produced by *P. fragi* including dimethyl sulfide, ethyl acetate, ethanol, methyl mercaptan, ethyl butyrate, and ethyl hexanoate.

2.3. Staphylococci and Micrococci

Gale (1962), in the course of a symposium devoted to staphylococci and micrococci, describes these organisms as having "a good round character." They are gram-positive, nonsporulating, nonmotile cocci; the staphylococci have fermentative and oxidative properties. Micrococci can only use glucose oxidatively, and this is one of the principal properties for separating the two genera. Staphylococci, on the other hand, grow well under microaerophilic conditions, utilizing glucose by a heterolactic fermentation, with pyruvate being a key intermediate

$$2 \text{ pyruvate} + \text{water} \longrightarrow \text{lactate} + \text{acetate} + \text{carbon dioxide}$$

It is probably this reaction which permits the incorporation of tricarboxylic-acid-cycle poisons into selective media. For example, the widely used Baird-Parker agar contains tellurite, and the obligately aerobic micrococci do not grow in this medium. The oxidative and fermentative pathways are readily uncoupled in staphylococci (Strasters and Winkler, 1963). An unusual feature of staphylococcal energy generation is that they can use arginine for the formation of ATP (Gale, 1962).

Although extensively studied, the taxonomy of these organisms remains muddled. The eighth edition of *Bergey's Manual* (Buchanan and Gibbons, 1974) recognizes three species of staphylococci, of which *S. aureus* is the coagulase-positive, pathogenic member; *Staphylococcus epidermidis* and *S. saprophyticus* are not generally regarded as pathogenic. In addition to the coagulase test, the thermostable nuclease test has most recently come into favor for distinguishing the pathogenic members of the genus (Rayman *et al.*, 1975). This enzyme persists in foods after a copious growth of the organisms dies out. This sometimes permits the detection of the organisms even when they are no longer present.

S. aureus occupies an important position in the hierarchy of the organisms of the food microbiologist; it is probably the most common cause of food-poisoning outbreaks in Western countries. This microbe produces a variety of toxins, of which the enterotoxins are responsible for food poisoning (Bergdoll, 1970). So far, seven enterotoxins have been described. They are all proteins and can be distinguished serologically. Types A, B, C_1 and C_2 have been particularly well studied, and in some cases their primary and secondary composition is known. They are heat stable and, at near-neutral pH values, largely survive cooking. Their amino acid composition or sequence gives no indication of why they should be so heat stable.

Enterotoxins A and C are most commonly responsible for food poisoning. However, all enterotoxins are maximally produced under optimal growth conditions, i.e., at 30–37°C in rich, well-aerated foods.

Why are enterotoxins produced? What is their ecological significance? Their synthesis resembles that of the bacilliary peptide antibiotics, and enterotoxins may be secondary metabolites. Unlike antibiotics, they do not appear to have antibacterial activity so that their synthesis does not confer an advantage to the producer organism. Peptide antibiotics inhibit the organism which produces them, so that they may be required in cellular differentiation (Katz and Demain, 1977). However, enterotoxin B appears to have no effect on the growth of the producer organism (Hurst and Kruse, 1972).

From time to time, reports appear indicating that growth and enterotoxin A synthesis can be dissociated (McCoy and Faber, 1966). Niskanen and Nurmi (1976) found that starter organisms in salami meat prevented growth and enterotoxin synthesis of type A but not type C.

It is generally accepted that about 10^6 cells/g of food are required for sufficient enterotoxin to be formed to constitute a health hazard. Staphylococci have the potential of reaching this concentration in a number of raw and fermented foods (salami-type sausage and cheeses). If for some reason, the starter develops slowly but the food is kept warm, an initial relatively low contamination of 10^3/g can reach 10^6/g in only 5 hr. This calculation assumes a 30-min doubling time; even more rapid growth has been recorded in batch-culture studies (Hurst et al., 1974). The high populations of staphylococci reached in fermented foods can only occur during the lag period of the LAB. Once the LAB develop, the staphylococci die off, and populations of 10^6/g or higher can disappear within a few days. Such foods may contain important quantities of enterotoxin without containing significant numbers of staphylococci.

Enterotoxin B differs from enterotoxin A in that the former is almost certainly plasmid controlled and may be linked with tetracycline resistance (Shalita et al., 1977). Its synthesis can be easily dissociated from growth. In foods its expression is readily inhibited by fermentable carbohydrates, possibly by catabolite repression (Morse and Mah, 1973), so that in practice it is rarely encountered as an etiological agent of food poisoning. Dornbusch and Hallander (1973) believe that enterotoxin B is important in human disease. The antibiotic-resistant strains they isolated in hospitals were generally strong producers of enterotoxin B.

The nature of the flora occurring in association with staphylococci may also determine the expression of enterotoxins. Gram-negative organisms almost always get the better of staphylococci. Meers (1973), quoting the work of Oberhofer and Frazier (1961), showed that although *E. coli* and *S. aureus* grew to similar cell densities when in axenic culture, *E. coli* outgrew *S. aureus* in mixed culture. This was because *E. coli* had a shorter generation time. Although it was not determined, it is likely that in such a situation *S. aureus* would be prevented from producing enterotoxin. Troller and Frazier (1963) concentrated an antibiotic substance from *E. coli* which was especially effective against staphylococci and micrococci. Collins-Thompson *et al.* (1973) grew *P. aeruginosa* in axenic and in mixed culture with an enterotoxin-B-producing strain of *S. aureus*. *P. fluorescens* "injured" *S. aureus*, as demonstrated by the decline in the viable count and by the loss of salt tolerance by the latter organism (Fig. 1). At the same time, the synthesis of enterotoxin B was also inhibited. The injury of *S. aureus* was not caused by medium depletion, lack of O_2, or pigment

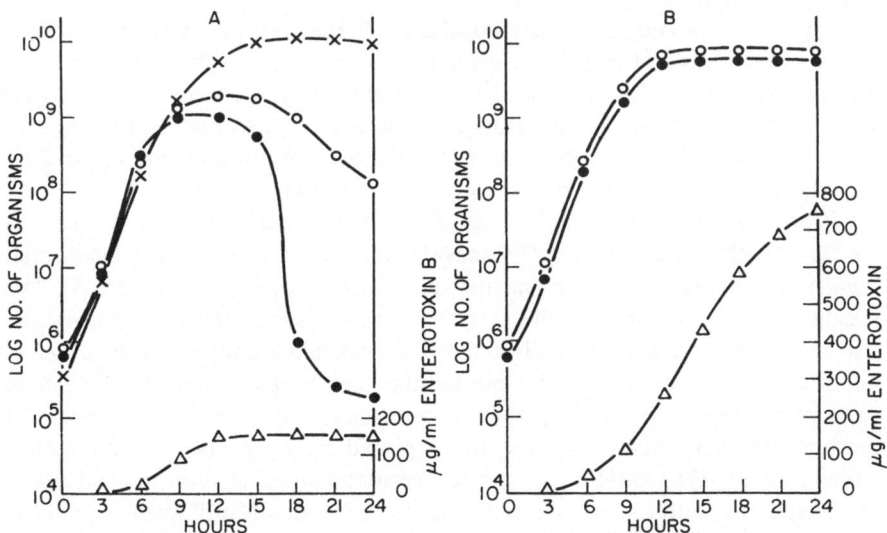

Figure 1. Interaction and injury of *Staphylococcus aureus* by *Pseudomonas aeruginosa*. A: The two organisms in mixed culture. B: *S. aureus* in pure culture. (×), Total count; (○), *S. aureus* counted on optimal medium; (●), *S. aureus* counted on stress medium (optimal medium containing 7% sodium chloride); (△), staphylococcal enterotoxin B (μg/ml of medium). After Collins-Thompson *et al.* (1973).

production but rather by an extracellular enzyme(s) which appeared to affect the integrity of the staphylococcal wall and/or membrane.

However, staphylococci do not appear to compete successfully even against other gram-positive bacteria. Oberhofer and Frazier (1961) reported that cultures of fecal streptococci inhibited *S. aureus*. Iandolo *et al.* (1965) observed the repression of *S. aureus* by *Streptococcus lactis* var. *diacetylactis*. The inhibition was ascribed to competition for nutrients; nicotinamide became unavailable to the staphylococci, especially at low pH. Kao and Frazier (1966) and Haines and Harmon (1973) made rather similar observations, though they used a much wider variety of competing microbes.

S. aureus can be regarded as an animal-skin organism. In man, the principal source is the nose (Defigueiredo and Splittstoesser, 1976). Estimates of the carrier rate of populations vary between 20% and 80%. From the nose, staphylococci can spread to other parts of the body, principally to different parts of the skin and hair. *S. aureus* "animal types" also occur commonly on the teats and udder of cows; pigs and poultry also harbor them. Staphylococci and micrococci also occur on beef hides. The hide microflora is to some extent dependent on the microflora of soils from the pastures on which the cattle graze. The microflora of the beef (meat) surfaces, in turn, is similar to that of the hides (Empey and Scott, 1939).

Staphylococci and some micrococci are salt tolerant, possibly because they are skin organisms and may be growing in the presence of salty skin secretions. The salt tolerance of staphylococci means that they grow in foods at lower water activity than any other pathogenic bacterium. The exact limiting a_w depends on the redox potential and other factors. Aerobically, the a_w is 0.85 and anaerobically, 0.90 (Nickerson and Sinskey, 1974).

Because food is often handled, the flora of the human skin is important to the flora of the food. Fortunately, coagulase-positive *S. aureus* is rare on normal human skin, *S. epidermidis* being the dominant member (Marples, 1965). The reason for the dominance of the coagulase-negative species over the pathogenic one is not clear, and it is possible that the coagulase-negative species plays an important and useful role in preventing the establishment of the more pathogenic species. For example, *S. aureus* readily colonizes the sterile skin of newborn infants but is subsequently replaced by coagulase-negative species (Meers, 1973). This could be due to competition and exhaustion of nutrients or production of antibiotics. A number of staphylococcal antibiotics have been isolated and partly purified, for example, by Barrow (1963), who described a polypeptide antibiotic, and by Hsu and Wiseman (1971), who describe substances named epidermidins. These were effective against a wide range of gram-positive organisms, other staphylococci being especially sensitive. The antibiotics contain no lipid, nucleic acid, carbohydrate, halogen, or protein. They were found to be dialyzable and resistant to heat and proteolytic enzymes.

Drying and dry storage of staphylococci in foods have been investigated by a number of workers and in some cases contrasted with the behavior of salmonellae. Lee *et al.* (1955) showed that *S. aureus* grew well in a lump of wet dough. However, after the population reached 10^9/g, *S. aureus* declined and became nonviable after 6 months of storage at $25°C$. In contrast, cells of *Salmonella typhimurium* were recovered after one year of storage. Similar results were obtained when staphylococci were inoculated into spaghetti dough (Walsh and Funke, 1975). Injured staphylococci in milk powder stored at $20°C$ declined from 10^7 to 10^2/g in 6 weeks (Hurst *et al.*, 1976).

This brief survey of the appearance and disappearance of staphylococci in foods and in ecological niches serves to emphasize that these organisms may have a good round character but are only moderately successful competitors. For this reason, the existence of large numbers of staphylococci in foods casts doubt on the practices associated with the production of that food. Because of their ubiquitous presence, small numbers of staphylococci should be expected in raw foods. After heat processing however, staphylococci should be absent (though in practice this occurs only in canned foods). Large numbers in heat-processed foods represent a potential hazard, and they are a certain indicator of unsatisfactory manufacturing or storage conditions after heating.

In contrast to staphylococci, micrococci are much less studied, probably because they are nonpathogenic, do not form enterotoxins, and, being biochemically more inert, are not so important as spoilage organisms. In certain foods, they form a substantial proportion of the flora at spoilage time. For example, Cavett (1962) reported that in vacuum-packed bacon stored at $20°C$, micrococci dominated over LAB. However, because of their biochemical inertness, the resulting spoilage was described as "scented-sour." When stored at $30°C$, the micrococci were displaced by the more biochemically active coagulase-negative staphylococci, which spoiled the product with a putrid odor. Kitchell (1962), discussing the micrococci of meat products, quotes earlier work of 1922: Micrococci were inoculated into experimental cans of corned beef and "produced no deleterious changes though growth occurred."

There are surprisingly frequent references to the presence of micrococci on fish, although they are not implicated in fish spoilage. One anticipates that micrococci would be found on salt-fish products, but they also appear to be part of the normal flora of fresh Pacific salmon (Snow and Beard, 1939) and other fish (Colwell, 1962; Pelroy *et al.*, 1967).

Like the staphylococci, micrococci produce antibiotics. Micrococcin, which was isolated by Su (1948), has the distinction of being one of the few characterized and tested substances. Niinivaara (1955) used a strain of antibiotic-producing *Microoccus* (M_{53}) for starter in fermented sausage (salami) manufacture. His M_{53} culture grew in the sausage meat and produced the antibiotic which inhibited the miscellaneous and undesirable organisms present. The LAB appeared to be antibiotic resistant and developed normally. The micrococci tended to disappear during the smoking of the sausage. Sausage factories using the micrococcus starter reported fewer defective products and improved flavor (Niinivaara and Pohja, 1957).

2.4. Lactic Acid Bacteria

2.4.1. General

In food microbiology, the term lactic acid bacteria (LAB) is frequently used but seldom defined. The concept of the group as it exists today is based essentially on the classical monograph of Orla-Jensen (1919), and it includes the genera *Lactobacillus, Leuconostoc, Pediococcus,* and *Streptococcus* (Ingram, 1975). Orla-Jensen's broad definition that they are "immotile, sporeless, grampositive cocci and rods, which in fermenting sugar form chiefly lactic acid" still holds today.

LAB are widely accepted as not possessing catalase, and it is interesting to note that Orla-Jensen's definition makes no mention of this enzyme or other porphyrin enzymes. Pediococci, however, are known to contain catalase (Felton *et al.*, 1953), and they are considered a part of the group. *Streptococcus faecalis* can be regarded almost as a founder member of the group, yet this organism also is known to produce catalase. In addition to reports that *S. faecalis* may contain catalase, other aerobic properties have been reported for this organism. For example, Whittenbury (1978) has shown that the yield of aerobically grown *S. faecalis* was significantly higher than that of cultures growing anaerobically.

Thus, catalase content does not exclude genera from membership in the LAB. *Microbacterium* was originally created by Orla-Jensen within the LAB, and it includes *M. thermosphactum*, which is weakly catalase positive and has many characters closely resembling streptococci (e.g., guanine and cytosine value of 36 mol/100 ml; Collins-Thompson *et al.*, 1972). It is one of the commonest organisms in raw meat, another property it shares with LAB. Its present classification is uncertain, and the genus has been temporarily placed among the coryneform group (Buchanan and Gibbons, 1974).

The whole of the genus *Lactobacillus* is usually thought of as belonging to the LAB. This genus is essentially unchanged from that proposed by Orla-Jensen (1919), except that a single genus now covers the original three genera proposed. Thus, the following terms have disappeared: *Thermobacterium* (for homofermentative thermophilic lactobacilli), *Betabacterium* (for heterofermentative thermophilic lactobacilli), and *Streptobacterium* (homofermentative mesophilic lactobacilli).

An organism, originally isolated from nursing infants, was named by Orla-Jensen (1919) *Thermobacterium bifidum.* Later, in the 1930s, considerable controversy and confusion developed around this organism. At present, it is not classified among lactobacilli but is a member of a distinct genus (*Bifidobacterium*) in the family of Actinomycetaceae. The current usage of the term LAB does not include this organism.

Aerococci should also be mentioned as distantly related to streptococci. As regards the gram-positive cocci, all five species of the genus *Pediococcus* and the six species of *Leuconostoc* are usually regarded as part of the LAB. The genus *Streptococcus* needs qualifying because the genus includes frank pathogens of animals and humans, organisms suspected of being pathogenic, and organisms associated with dental caries. These are also called the pyogenic, *viridans*, and miscellaneous streptococci and are excluded from the LAB. The enterococci (fecal streptococci, serological group D) and the lactic streptococci (serological group N) are included in the term LAB and are among the principal constituent organisms. An important member of the lactic streptococci is the thermophilic *Streptococcus thermophilus*, which lacks the grouping antigen. Figure 2 is a graphic representation of the LAB concept.

Having discussed what we mean by LAB, we should discuss their activities in foods. Their most important characteristic is their competitiveness; in numerous situations, they become the dominant flora in practically pure culture. The possible reasons for dominance are, firstly, that they are expert at producing sub-

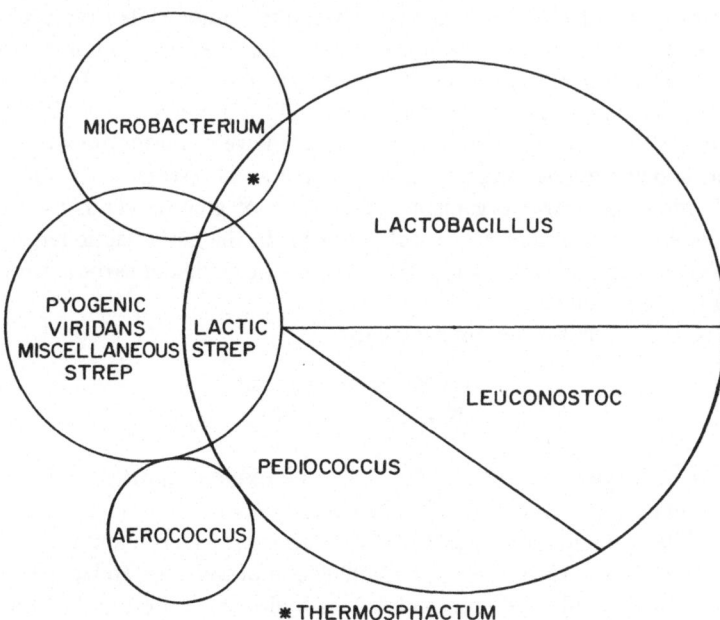

Figure 2. A model of the concept of lactic acid bacteria.

stances inhibitory to other microbes and, secondly, that they tolerate the adverse conditions of their own making better than most other microbes. This latter statement is made with raw, spoiling foods in mind; it is evident that spore-forming microbes have a higher heat resistance than any of the LAB.

2.4.2. Lactic Acid Bacteria in Competition with Other Microbes

2.4.2a. Limiting Acidity. LAB ferment many carbohydrates to lactic acid (homofermentative) or lactic acid and other products (heterofermentative). Although LAB are fastidious and may be difficult to grow in artificial culture, they grow rapidly in foods and lower the pH to a point where other organisms can neither initiate growth nor sustain growth which has already started. This explains how lactic acid, which contains about 90% of the energy of the glucose from which it is derived, accumulates in the presence of organisms which can potentially use it. In artificial culture, the pH may become so low as to become self-limiting. Some of the lactobacilli and pediococci are the most powerful acid producers, reaching a self-limiting pH of about 3.5. Lactic streptococci and leuconostocs have a limiting pH of 4.0–4.5. In contrast, gram-negative spoilage organisms such as the pseudomonads and other common spoilage organisms, in general, require a much higher pH for initiation and maintenance of growth, e.g., 5.5 or higher.

Ingram *et al.* (1956) showed that it was the undissociated molecule of the organic acid which was the food preservative. The accepted opinion is that lactic acid per se is a poor inhibitor and is ineffective above pH 5.5. Acetic acid (and its salts) are much more effective inhibitors and food preservatives. Organic acids are readily soluble in cell membranes and are believed to interfere with substrate transport and energy-yielding processes (Cramer and Prestegard, 1977).

In homofermentative organisms, each mole of glucose yields two moles of lactic acid by the Embden–Meyerhof pathway. In the heterolactic fermentation, one molecule each of lactic acid, ethanol or acetic acid, and carbon dioxide constitute the end products.

Pyruvate is usually a key intermediate:

$$\text{pyruvate} \rightleftharpoons \text{lactate}$$
$$\text{NADH} \qquad \text{NAD}$$

2.4.2b. Hydrogen Peroxide. Hydrogen peroxide is another end product of the fermentative activity of LAB and contributes to their successful competitiveness. The problem these organisms face is to supply nicotinamide dinucleotide in the reduced form (NADH) for the conversion of pyruvate to lactate. In higher organisms, lactic acid itself is reoxidized via hydrogen carriers and is thus further utilized. In LAB, lactic acid is the end product of their metabolism, and it is in the D, L, or DL optical configuration. Other microbes, e.g., enterobacteria, can use lactate as an energy source. The accumulation of lactate in some spoiling foods made it seem possible that other microbes failed to utilize it because of its optical configuration. However, Bennett *et al.* (1966) found that the optical activity of the lactic acid had little or no effect on its utilization by enterobacteria. The latter organisms, although they do not appear to contain a lactic racemase, did contain separate D- and L-lactate dehydrogenases.

In LAB, the reduced NAD is oxidized by flavin nucleotides, which react rapidly with gaseous O_2, leading to the formation of hydrogen peroxide.

$$\text{flavinH}_2 + O_2 \longrightarrow \text{flavin} + \text{hydrogen peroxide}$$

Flavin nucleotides are said to be autooxidizable. In the microbial cell, they occur associated with proteins; such flavoproteins may themselves have antibacterial properties because they generate hydrogen peroxide (e.g., glucose oxidase).

The powerful antibacterial property of hydrogen peroxide has been known for a long time. Its potency under "competitive" conditions can be illustrated by the fact that glucose oxidase was mistaken for an antibiotic (notatin, penicillin B) in culture fluids of *Penicillium* sp. (Coulthard *et al.*, 1945). Another fact known for a long time, and more recently overlooked, is that LAB which produce peroxide are more resistant to it than many other organisms. Wheater *et al.* (1952) showed that when about 10^4 cells/ml were used as an inoculum, the minimal inhibitory concentration (MIC) of hydrogen peroxide to *Lactobacillus lactis* was 125 μg/ml, whereas *S. aureus* had an MIC of 4 μg/ml. In these tests, the low level of cells used for inoculum ensured that insufficient preformed catalase was carried over into the MIC tubes. The *S. aureus* cells thus depended for their protection against hydrogen peroxide on their intrinsic cellular structure rather than on preformed catalase, a situation similar to *L. lactis*, which has no catalase. Under these testing conditions, there was a 30-fold difference in the sensitivity of the organisms.

In the manufacture of Swiss-type cheese, hydrogen peroxide probably plays a very important role in inhibiting the development of spoilage clostridia (Hirsch *et al.*, 1952). This is a good example of interaction in food: It appears that peroxide may be produced in the interior of large cheeses where the amount of gaseous O_2 is quite low. (This subject is further discussed below in Section 3.6.)

2.4.2c. Resistance to Carbon Dioxide. Packaging of foods, especially meat, radically alters the composition of the bacterial flora. Packaging therefore has marked ecological consequences, but it is beyond the scope of this article to deal with packaging per se. Containers are either rigid (cans and glass) and completely impervious to gases or flexible (plastics and foil); the latter may be pervious to different gases (for review, see Cavett, 1968; Briston, 1976). By various combinations of materials, it is possible to produce plastic packages having a wide variety of properties.

When a raw food (i.e., meat) or a food containing an active bacterial population (i.e., cheese) is packed, available O_2 can be rapidly exhausted within the pack, and carbon dioxide accumulates. The antibacterial properties of carbon dioxide have been known since 1933 (reviewed by Clark and Lentz, 1969b). Psychrophilic, aerobic gram-negative spoilage organisms are especially sensitive. For example, Alm et al. (1961) examined sliced processed-meat products that were vacuum packed in cellophane-polyethylene bags. The microflora changed from a mixed population of Bacillus, Achromobacter, and Lactobacillus to an almost pure culture of Lactobacillus. Similar results are reported by many other workers [e.g., Clark and Lentz (1969b); Kempton and Bobier (1970)]. It is interesting to note that the latter workers found that cooked ham, which contained much less carbohydrate than the other products examined, remained at above pH 6.0 but did not spoil by putrefaction. It seems that the effect of carbon dioxide is separate and additional to that of pH. The resistance of LAB to carbon dioxide, rather than the absence of O_2, may be one of the most important factors causing their dominance in packaged foods (Kitchell and Shaw, 1975).

2.4.2d. Antibiotic Production. S. lactis produces the antibiotic nisin, which is used extensively in many countries as a food preservative. The antibiotic may be formed in farmhouse cheese. Diplococcin is the name of another antibiotic produced by *Streptococcus cremoris*. Both antibiotics have been concentrated; the chemical and biological properties of nisin have been studied extensively. Nisin is further discussed below in Section 3.6 (for review, see Hurst, 1973, 1978). On the other hand, the antibiotics produced by lactobacilli have been studied less intensively, and their very existence is in doubt. Several "antibiotics" were probably artifacts due to hydrogen peroxide or low pH (Wheater et al., 1952). Vincent et al. (1959) concentrated an antibacterial factor from cultures of *Lactobacillus acidophilus*; there appears to be no further news of this antibiotic. Reuter (1971) believed that the antibacterial effect of lactobacilli in fermented sausages could not be explained in terms of peroxide or acidity alone. However, so far, he has not reported the isolation of an antibiotic from sausage lactobacilli.

Nisin and diplococcin are two polypeptide antibiotics produced by lactic streptococci. They are only effective against gram-positive organisms. However, in conjunction with other preservative techniques, these antibiotics probably contribute toward ensuring the dominance of these strains under some circumstances. It is possible that other antibiotics are also produced by other members of the LAB; however, convincing evidence for their existence is lacking.

2.4.3. Evolution of the Lactic Streptococci

Two species, *S. lactis* and *S. cremoris*, constitute the lactic streptococci. According to Hirsch (1952), these organisms are of relatively recent origin, now adapting to spilled milk. They are in deadly combat for dominance in this relatively recent environment. In other words, domestication of milk-producing animals and the production of milk under primitive conditions have created a new ecological niche for which these two species are combating. The arguments may be summarized as follows: Strains of both organisms produce antibiotics most effective against the other organism; e.g., *S. cremoris* is the most sensitive organism to nisin produced by *S. lactis*. Neither organism is pathogenic, and both prefer to grow in milk at room temperature and not at 37°C in the cow. In addition, although both organisms ferment lactose vigorously, they can both easily lose this property. Hirsch (1952) argues that it is surprising that lactose fermentation is not a more stable character in organisms specializing in milk souring. In partial support of this argument, Cords *et al.* (1974) showed that β-galactosidase synthesis in *S. lactis* was episomally controlled.

2.4.4. Other Factors

LAB are fastidious feeders, but, nevertheless, they compete well in many foods: meats and dairy and vegetable products. These all contain readily fermentable carbohydrates, growth factors, and acceptable nitrogen sources. Since many of these organisms have complex nutritional requirements, it is reasonable to suggest that commensalism and mutualism play an important part in regulating their populations.

Resistance of LAB to sodium chloride and low O_2 is not remarkable, but neither are they as sensitive as the pseudomonads and enterobacteria. For example, in brine-cured meats, pseudomonads are excluded by about 5% sodium chloride, while the growth of lactobacilli is favored (Bartl, 1973). In cucumber fermentations, LAB develop rapidly at 2–5% salt (Pederson and Ward, 1949). The actual salt concentration at which growth is prevented is much higher; an absolute figure is almost meaningless since this depends on pH, a_w, and the temperature at which the test is done (Ingram and Mackey, 1976).

Similarly, minimum levels of water activity (a_w) permitting growth are interdependent with other physical factors; LAB are about as resistant as gram-negative organisms. The limiting a_w for *E. coli* and *Lactobacillus viridescens* is 0.95 (Troller and Christian, 1978).

In pasteurized hams heated to an internal temperature of only about 68°C, enterococci (*S. faecalis* and *Streptococcus faecium*) may survive and may become the major spoilage organisms. Although reducing a_w (by increasing the sodium chloride concentration, for example) may improve the stability of the product, this is inadvisable because the heat resistance of enterococci might increase with decreasing a_w. The heat resistance of lactobacilli also increases with decreasing a_w (Leistner and Rödel, 1976a), and this may result in decreased shelf life.

2.4.5. Competition among Lactic Acid Bacteria

2.4.5a. Starters. Starters are widely used industrially for the preparation of fermented foods. Cheese starters are probably the most widely studied, and (except in New Zealand) they contain several strains of closely allied lactic streptococci. These strains are distinguishable by their sensitivity to bacteriophage attack. The composition of these "multiple-strain" starters changes with subculturing, one of the strains becoming dominant. The cause of dominance is not fully understood and may be a combination of antibiotic production and competitive growth rate. Reddy *et al.* (1971) observed that, in mixed cultures in milk, *S. cremoris* tended to dominate *S. lactis*. Strains of both of these organisms produce antibiotics which are most effective against the other organism.

2.4.5b. Successions. The following succession often occurs in foods:

$$\text{mixed flora} \longrightarrow \text{lactic cocci} \longrightarrow \text{lactobacilli}$$

This course of events can be unfailingly observed in cheese made from raw milk and inoculated with a starter of lactic streptococci. Within a few days, the initial mixed flora is suppressed by the streptococci. These then reach 10^8–10^9/g of fresh cheese. (At this time, lactose disappears, either because it is utilized or because it is lost in the whey.) Thereafter, lactobacilli gradually take over and reach numbers comparable to those orginally attained by the streptococci. A mature Cheddar cheese is likely to contain very few lactic streptococci, 10^6–10^7 fecal streptococci/g and 10^8–10^9 lactobacilli (streptobacteria)/g. The cause of this succession is not established. It may be that lactobacilli are better able to grow than streptococci either on traces of carbohydrate (e.g., lactose-free fresh cheese, which may contain citrate) or on nitrogenous substrates. In the case of Swiss-type cheese, the succession is easier to understand: It is a high-temperature fermentation initiated by *S. thermophilus*, which is followed by thermophilic lactobacilli, somewhat as described below for yogurt. As the cheese cools, the streptobacteria replace the thermobacteria.

A similar sequence of events occurs in sauerkraut, and the process was carefully studied by Stamer (1975). He noted that *Leuconostoc mesenteroides* was the first dominant LAB, and this is usually terminated by the succeeding association of *L. plantarum* and *L. brevis*. Based on pure-culture studies done in filtered sauerkraut juice, Stamer interprets these results as follows: *L. mesenteroides* develops rapidly but has a short life span because of its sensitivity to the acids it produces. The lactobacilli, especially *L. brevis*, grow more slowly but they are more tolerant of the low pH.

The influence of sodium chloride concentration, the interaction with pH, and the effect on lag and generation times were also studied. The heterolactic fermenters were much more sensitive than the homolactic fermenters, with *Pediococcus cerevisiae* being perhaps the most salt-independent organism.

2.4.6. Mutual Stimulation

There are several reports of the growth stimulation of LAB grown in mixed culture (see Hurst, 1973). Peptides produced by one species are generally responsible. Yogurt is an associative culture of *S. thermophilus* and *Lactobacillus bulgaricus*. The two organisms are mutually beneficial, the streptococcus initiating growth which removes O_2, creates weakly acid conditions, and produces stimulating peptides. The lactobacillus then continues by further hydrolyzing lactose and casein (Davis, 1975). Another interesting example is the association of LAB with yeasts which occurs in certain fermented milk, sourdough bread, and Roquefort cheese. In the last case, it is known that the lactic streptococci and *Leuconostoc* produce acid and gas more vigorously when they are cultivated along with yeasts. The yeasts may be supplemented or replaced by proteolytic fecal streptococci. It appears that the good quality of Roquefort cheese may result from a satisfactory balance between these different microorganisms (Mocquot, 1971).

2.5. Spore-Forming Bacteria

The two genera of this family of major importance to food microbiology are the aerobic *Bacillus* and the anaerobic *Clostridium*. These organisms are not particularly successful competitors, and they are seldom, if ever, dominant in foods which have not been heated. Their ability to form spores, however, ensures that they can survive many unfavorable physical treatments. Sporulation, thus viewed, is not a mechanism designed for dominance but for survival. When more favorable conditions return, spores germinate and outgrow; for example, in spoiling canned foods, these organisms can be the dominant flora. Thus, the basis of all canning process calculations is the prevention of *Clostridium botulinum* spores from surviving the processing.

It is interesting to note in passing that spore germination is never complete, and a variable fraction called superdormant remains ungerminated (Gould, 1969).

It is as though some venturesome individuals take a look at the outside world while others wait within the safety of their resistant structures. This mechanism assures that some individuals always survive even if there is an unexpected return to unfavorable physical conditions.

The lack of competitiveness of these organisms in unheated foods is possibly due to their slower growth rate and the apparent lack of formation of antagonistic substances. For example, they do not significantly lower the pH or produce hydrogen peroxide. Antibiotic production is an important and widespread property among the bacilli (it does not appear to occur among clostridia), but antibiotic synthesis occurs late in the growth cycle (Katz and Demain, 1977) and may not help the competitiveness of the organism in foods. The very absence of dominance of bacilli in unheated foods suggests that antibiotics may not be formed.

C. botulinum cannot develop below pH 4.5 and it is on this basis that canned foods are classified as high-acid, which require mild heating only to destroy spoilage organisms, since *C. botulinum* cannot grow and produce its toxins, and low-acid foods (above pH 4.5), which require sufficient heat to inactivate *C. botulinum*.

Another reason for their lack of success as competitors is that most spore formers do not develop at below 10°C, with the important exception of type E *C. botulinum*, which can develop down to about 5°C. Spore formers are classified in Table II according to their resistance to acidity and their temperature requirements. Those at the top of the table are the more heat-resistant ones; there is a broad correlation between the temperature for growth (formation of vegetative cells) and thermal resistance of spores. Spores of these organisms are

Table II. Spore-Forming Bacteria Important in Food Bacteriology[a]

Approx. temp range for good growth	Acidity status of food	
	"Acid" (pH 4.0–4.5)	"Low-acid" (pH > 4.5)
Thermophilic 35–55°C	*Bacillus thermoacidurans*	*Clostridium thermosaccharolyticum* *Clostridium perfringens* *Bacillus stearothermophilus*
Mesophilic 10–40°C	*Clostridium butyricum* *Clostridium pasteurianum* *Bacillus macerans* *Bacillus polymyxa*	*Clostridium botulinum* A, B *Clostridium sporogenes* *Bacillus subtilis* *Bacillus cereus*
Cold-tolerant > 5°C		*Clostridium botulinum* E

[a]Modified from Ingram (1969).

frequently present in canned foods, which nevertheless remain stable indefinitely in temperate climates. Storage at higher temperatures may permit these spores to outgrow and spoil the food. Organisms listed under the heading "acid" are nonpathogenic spoilage organisms. They may be troublesome in canned fruits since they may survive heating and grow in the acid environment. The organisms listed under "low-acid" include both spoilage and pathogenic types. The most important of all is *C. botulinum*. Without exception, all food processes have to consider the possibility of this organism developing. The process must be so designed that development is prevented with an extremely high degree of probability.

The term "botulinum" was first used in Germany during the late 19th century to denote an organism from sausage. However, its connection with meat products has been appreciated for about 1000 years; its history was described by Dolman (1964). Seven serological types of the toxins are known, of which types A and B most commonly affect humans and types C and D may be regarded as animal types. Type E affects man but is most commonly isolated from fish. Proteolytic and nonproteolytic strains exist in most types (Hobbs, 1976).

Intensive surveys since 1922 showed that *C. botulinum* types A-D are primarily soil organisms. Even the type E in fish appears to be of soil origin. Riemann (1973) reviewed the evidence in a perceptive and well-informed article. Type E organisms occur with an unusually high incidence in the Baltic Sea. Johannsen (1965) explains this by the fact that this sea receives drainage from a land area almost four times larger than the sea area. Further direct support for the assumption that the distribution of type E spores depends on sedimentation comes from the work of Laycock and Loring (1972). They found a direct correlation between type E distribution and terrigenous sedimentation in the Gulf of St. Lawrence. Type E also occurs in freshwater fish and sediment, and Lake Michigan has been especially well studied. Multiplication in vegetation from freshwater has been shown to occur probably because of the animal debris associated with the vegetation. Riemann (1973) suggests that the ecological niche of *C. botulinum* remains unknown; type E has not the properties of a marine or freshwater organism, and it is likely that spores of the organism survive in nature to multiply only in vertebrate and invertebrate carrion. Therefore, the general view is that *C. botulinum*, although not a normal intestinal organism, can be ingested and carried in the spore form. The most common vertebrate carrier is fish. Invertebrates are completely resistant to botulinal toxins and can pick up and carry spores by feeding on contaminated carrion. Animal corpses of many types may carry the organism, e.g., corpses of dogs, cats, rats, mice, and poultry.

Waterfowl are sensitive to type C botulinum and may be killed in large numbers. Riemann (1973) suggests that "the epidemic is perpetuated mainly through a duck–maggot cycle. The birds which consume toxic maggots become a medium for *C. botulinum* after death and a new crop of toxic maggots is produced."

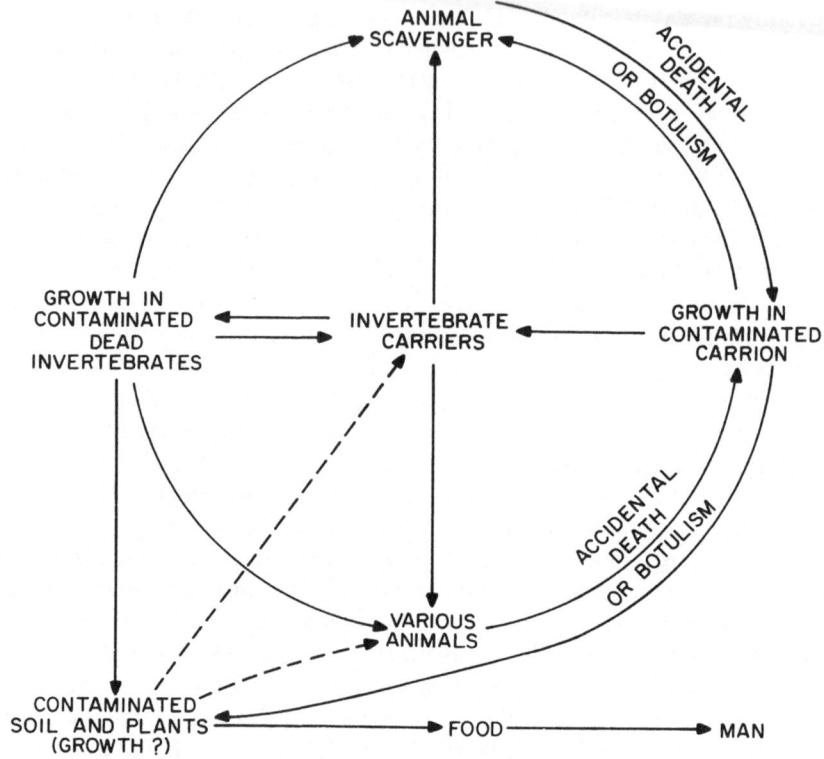

Figure 3. Suggested cycle of *Clostridium botulinum* types A, B, C, or D in nature. After Riemann (1973); reprinted by permission of the author.

    Figures 3 and 4 are hypothetical cycles of botulism in land and water environments, respectively. Figure 3 might illustrate the occurrence of types A–D; Fig. 4 refers to type E. Man may be an accidental victim but plays an insignificant role in these cycles.

    Two aspects remain puzzling regarding the distribution of type E *C. botulinum*. The first one concerns its virtual absence from the waters close to the British Isles, although it occurs in the North Sea. The second point is the apparent absence of the organism from the southern hemisphere. Other types have been reported from South Africa but not type E (Meyer, 1956).

    When *C. botulinum* develops, toxin is formed in the food, and unless the toxin is destroyed (e.g., by proteolytic enzymes from other microorganisms), the food can be lethal to humans. *Clostridium perfringens* is another food-poisoning clostridium, although much less dangerous than the former organism, producing an illness that is relatively mild. It is ubiquitous, and among patho-

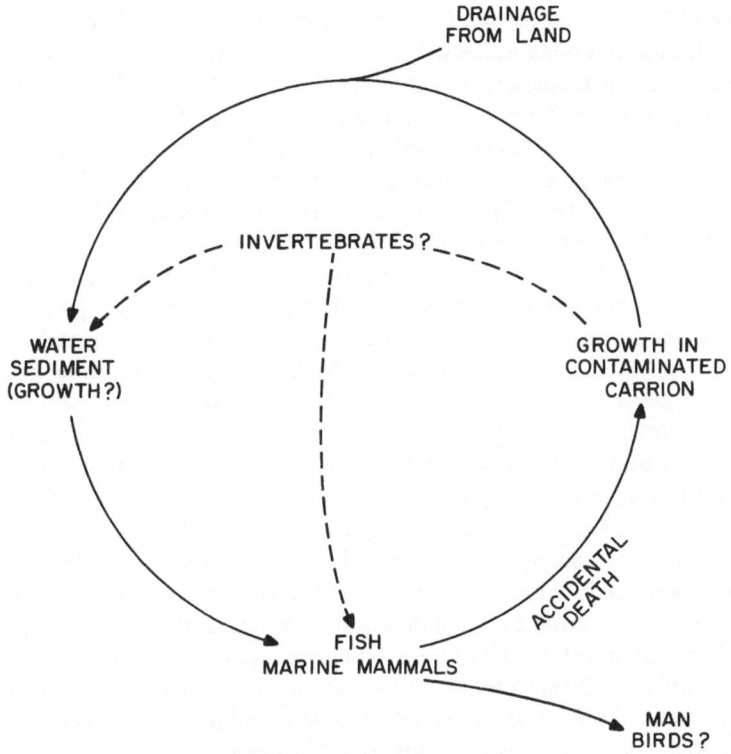

Figure 4. Suggested cycle of *Clostridium botulinum* type E in nature. After Riemann (1973); reprinted by permission of the author.

gens, it is possibly the most widely distributed in nature. It may be isolated from soil and the intestinal contents of many animals and most humans. Among humans, the carrier rate is probably close to 100% (Defigueiredo and Splittstoesser, 1976). The incidence of the organism is routinely higher in meat, stews, meat pies, and poultry products than in other foods. The organism does not produce a preformed toxin. Food poisoning is caused when large numbers of cells are ingested (e.g., 10^6–10^7/g) with a food in which the organism can grow. Enterotoxin is then formed in the bowel. A correlation has been demonstrated between sporulation and toxigenesis; nonsporulating vegetative cells do not produce enterotoxin (Hauschild, 1974).

Bacillus cereus is another potential food-poisoning organism listed in Table II. It is commonly and justifiably considered to be a harmless saprophyte under most circumstances (Goepfert *et al.*, 1972). Special circumstances are required for it to become a food-poisoning organism. This occurred in the cases studied by Hauge (1955), which are the first classical descriptions of this gastroenteritis. Large quantities of vanilla sauce were prepared the day before its consumption and stored at room temperature. The preparation of the sauce killed competing flora, and the slow cooling of the large bulk of the food enabled the spores to develop and to reach numbers of more than 10^7/g. The organism is an important pathogen in Hungary, and it is said that this may be due to the highly spiced meat dishes popular there. Spices often contain large numbers of aerobic spore formers (Kim and Goepfert, 1971; Goepfert *et al.*, 1972).

B. cereus is the only aerobic spore former which is known to be the occasional cause of mild food poisoning. In contrast, *C. botulinum* produces the most powerful known toxin, so that one might wonder at the rationale for the enormous increase of vacuum packing of foods which has occurred in recent years. This question was raised following the incrimination of vacuum-packed smoked fish in outbreaks of botulism. For example, it was suggested that *C. botulinum* might grow better in a vacuum than in non-vacuum-packed food. Lee and Foster (1965) investigated this with sliced bologna sausage inoculated with spores. They found no difference in the rate of toxin development, presumably because conditions were already anaerobic between the meat slices. On the other hand, Sugiyama and Yang (1975) reported the enhancement of toxigenesis of the organism in mushrooms packed in semipermeable plastic film.

Another important consideration is the interaction of some food additives with *C. botulinum*. Perhaps the most important and controversial one is sodium nitrite ($NaNO_2$). This salt is extensively used in cured meats, to which it imparts a characteristic color, but its chief function is to assure the safety of the product. Sodium nitrite inhibits *C. botulinum*. Food-regulatory agencies would like to cut down on its use or eliminate it altogether because of the fear of cancer. Nitrites may form nitrosamines as a result of reaction with amines, amino acids, etc., and these can be converted to carcinogens in various organs. Nitrosamines have occasionally been detected in foods, but only at few parts per billion (Sen, 1974). However, so far no suitable substitute has been found for nitrite.

Nitrite disappears when heated in meat, but an inhibitor of unknown composition is formed (Ashworth and Spencer, 1972; Pivnick and Chang, 1973). It is thought that this inhibitor plays an important role in canned semipreserved cured meats in preventing growth of the small number of bacterial spores that survive the heating process.

3. The Major Food Products

3.1. Eggs

The egg shell and membrane resist penetration of the interior by bacteria. Nevertheless, work by Haines (1938) and others suggests that about 5% of newly laid eggs may be contaminated with bacteria. Infection of the egg contents can take place prior to the laying of the egg, by *Salmonella pullorum* and *S. gallinarum*, if the ovary of the bird is infected. This type of internal infection is more common with ducks' eggs than with hens'. In the early stages of egg holding, organisms like *Lactobacillus* and *Micrococcus* have also been reported to be within the shell (Harry, 1963).

The surface of the egg shell is contaminated from a number of sources. When the egg is being laid, it passes the cloaca, and as a result, extensive contamination with intestinal bacteria of the bird takes place. Contact with litter and nesting material on floors adds to the build-up of the bacterial flora on the shell. Handling by humans and manufacturing procedures such as washing can introduce pseudomonads and staphylococci. Thus, the flora of the shell is mixed, consisting chiefly of gram-negative organisms such as *Salmonella, Proteus, Escherichia, Pseudomonas*, and other coliform bacteria such as *Citrobacter*. The major gram-positive groups include *Streptococcus* and *Micrococcus*. When the shell dries out, organisms like the micrococcus commonly dominate because of their resistance to desiccation. The external structures of the shell (mucin layer, shell, and the membranes inside the shell) are capable of supporting bacterial growth, especially of pseudomonads and salmonellae. The importance of the surface contamination lies in the technology of producing liquid egg. It is difficult to free the shells of the gram-negative flora, and until recently liquid egg was quite a hazardous, though widely used, product. Liquid egg is now pasteurized.

Bacteria penetrate the shell through the pores when the mucin layer that covers the shell is removed by washing or is missing. Organisms on the shells of newly laid warm eggs can be drawn through the pores as the egg cools. Once bacteria enter the egg, invasion of the albumen and yolk can take place. Because of their predominance on the shell, these bacteria tend to be gram negative. It is thought that their entry is further facilitated by the fact that they are usually motile.

Not all organisms have the same ability to penetrate the shell. Lifshitz *et al.* (1964, 1965) showed that organisms such as *P. fluorescens* were capable of penetrating the shell structure more quickly than salmonellae. They also showed that the resistance to penetration of the outer membrane of the shell by *P. fluorescens* was much less than that of the shell or of the inner membrane.

Physical factors encouraging the penetration of the shell by bacteria are defective shells (cracks), lack of mucin layers, or high storage temperature. However, the most important factor is moisture (Williams *et al.*, 1968).

The type of flora which invades the egg contents is usually a mixture of gram-negative bacteria. This type of flora is least sensitive to both lysozyme and the alkaline pH ($\simeq$9.2) of the albumen (Board and Fuller, 1974). It is believed that these two factors are the principal defense mechanism of the inner portion of the egg.

There is little information about the type of successions of microorganisms that occurs when a mixed flora invades the yolk or albumen. The only clue as to dominance of one species over the others is the frequency with which different types of breakdown products occur in the egg. The most frequent organisms isolated from "rotting" eggs are the pseudomonads. It may be because most eggs destined for the market are kept at temperatures below 15°C, a temperature favoring such organisms. *Proteus, Alcaligenes, Aeromonas, Citrobacter,* and *Serratia* all appear to have an equal chance to thrive and survive once they penetrate into the egg contents (Board, 1970).

3.2. Fish and Shellfish

Fish and shellfish are among the few foods in which the state of the environment is reflected by the microflora of the product. The sedentary way of life of oysters, for example, causes them to have a microflora which reflects the microbiology of the surrounding waters. The finding of *E. coli* in oysters is an indication of fecal pollution in the water from which the oysters were taken.

Similarly, the microflora of fish also varies quantitatively and qualitatively, depending on the state of the water from which the fish was caught. The composition of the microbial flora of the skin depends on its origin (fresh- or saltwater fish) and on the climatic conditions. Studies by Shewan (1962) have shown that the bacteria associated with fish are usually distributed on the skin, gills, and intestines. The internal flesh is usually uncontaminated. The levels of bacteria on skin or gills range from 10^2 to $10^8/cm^2$, depending on the water quality. Similar levels have also been reported in shellfish.

Psychrotrophic types of bacteria are usually associated with fish and shellfish caught in temperate climates (Shaw and Shewan, 1968). Fish taken from warmer climates tend to support mesophilic bacteria. The differences between these two groups, however, are usually minimal once the fish is processed, since the chilling temperature selects out the psychrophiles.

The mixed flora found on fish caught from oceans in a temperate climate consists of the genera *Acinetobacter/Moraxella, Pseudomonas, Flavobacterium, Cytophaga,* and *Vibrio.* In tropical and subtropical climates, bacterial populations contain, in addition, a number of gram-positive genera such as *Bacillus, Micrococcus,* and coryneforms (Shewan, 1962; Gillespie and Macrae, 1975).

Freshwater fish have a mixed flora consisting of both gram-negative and gram-positive groups. In addition to those gram-negative bacteria already listed for fish caught in temperate waters, some members of the coliform group are also present, i.e., *E. coli,* probably from polluted water. Shellfish have a large mixed gram-negative flora similar to that of fish. *Bacillus* and *Micrococcus* are additional important parts of the flora (Vasconcelos and Lee, 1972). This microflora appears to be chiefly aerobic; however, facultative organisms (e.g., *Vibrio*) are also found.

After fish are caught and stored, the developing microflora is dominated by the genus *Pseudomonas*. This is essentially due to the temperature of storage. The species of pseudomonads of importance in fish spoilage are generally carbohydrate oxidizers but are variable pigment producers. Different species of fish show some variation in the microflora developing during storage. For example, Shewan (1971) found with various fish stored on ice that after three days the pseudomonads made up about 20% of the flora. After 15 days, 98% of the flora was accounted for by the pseudomonads. The other important genus, *Acineto-bacter/Moraxella*, changed during the 15-day storage from about 20% to 1%. Other groups, including *Flavobacterium* and *Micrococcus*, disappeared quite rapidly. On the other hand, in studies by Laycock and Regier (1970) with iced whole haddock, *Acinetobacter/Moraxella* appeared to compete well with the pseudomonads (Fig. 5). The pseudomonad population consisted of the pigmented and nonpigmented types.

The ability of the pseudomonads to usually dominate in fish and many other foods may be due, in part, to the shorter lag phase of growth of these organisms at low temperatures. The diversity of metabolic activity of this group also causes the greatest biochemical changes in fish. Shewan (1961) suggested that the pseudomonads utilize various simple compounds such as creatine, trimethyl-

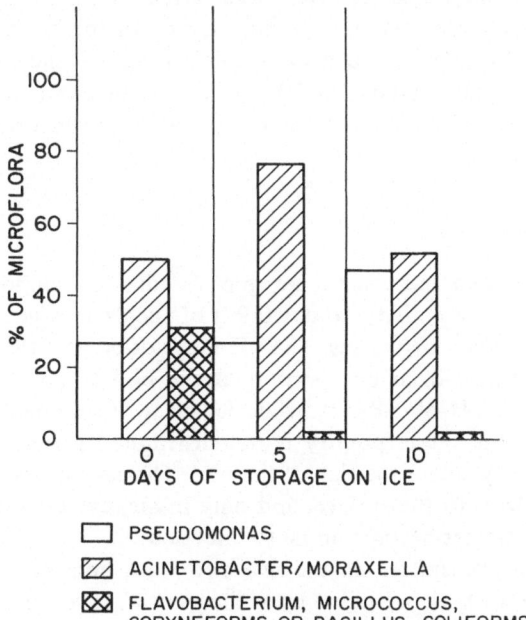

Figure 5. Histogram of the development of bacteria in fish kept on ice.

amine oxide, taurine, and amino acids at a greater rate than other microorganisms. The resulting compounds associated with breakdown of fish tissues include hydrogen sulfide, trimethylamine, indole, and ammonia (Lerke *et al.*, 1963, 1965). Pseudomonads are not wholly responsible for the biological breakdown, since fish is also prone to autolysis. The role of gram-positive organisms such as *Micrococcus* and *Bacillus*, which are often reported on fish, was so limited that they cannot be considered indigenous to fish. Work by Shewan and Hobbs (1967) and others suggests that *Micrococcus* species are associated with fish because of handling and that they are contaminants from human sources. This notion is supported by Harrison and Lee (1969) and Vanderzant *et al.* (1970), who worked with shrimp. The initial levels of *Micrococcus* species tend to be much higher than for fish, reflecting increased handling, but the microflora developing on stored shrimp is similar to that of fish.

Shellfish, such as clams and oysters, develop a different microflora during storage. The major reason for this is that shellfish have a higher glycogen content (3–5%) than fish (none). In shellfish stored at refrigeration temperatures, *Pseudomonas* and *Acinetobacter/Moraxella* tend to dominate in the early periods of storage. A more fermentative flora of lactobacilli and streptococci develops later, and they compete by lowering the pH of the clam or oyster to 5.5. LAB then dominate, especially if the storage temperature is around 10°C.

Vibrio parahaemolyticus deserves brief attention since this organism is unique to fish and shellfish and is considered part of the normal microflora of many species. Under refrigerated storage, this organism is incapable of competing well with the other bacteria. At 20°C or above, however, this organism not only grows well but is capable of becoming the dominant organism in fish or shellfish.

3.3. Poultry Meats

Poultry meats, like red meats, are good media for supporting microbial growth. The high water activity (about 0.99) of such meats permits the growth of a wide range of microorganisms. The pH range of poultry tissues, unlike that of carcass meat, varies from one portion of the bird to another. Breast-meat tissue usually has a pH of 5.8–5.9, while leg-meat tissue ranges from 6.4–6.7. Unlike red meat, the fat in poultry is not distributed throughout the tissues but is located mostly under the skin of the bird. The skin serves to limit most of the microbial activity to the surface, and only in areas where the skin is broken do the underlying tissues become prone to infection.

Organisms are transmitted to the live bird via a number of sources. Contamination begins with eggs. Studies by Williams *et al.* (1968) have indicated the possible penetration by *S. typhimurium* through the shell of eggs. Live birds can also be infected through contact with fecal contamination from feed, soil, water, and other infected birds or animals. These routes of transmission have been studied particularly with salmonellae in mind. Poultry is the largest single reservoir for salmonellae in the human environment (Anonymous, 1969). Fully processed poultry meat often contains *Salmonella*, from about 10 to 100/g of skin.

Surface contamination plays a major role in determining the various kinds of microbes which gain access. The levels and types of flora are further influenced by the preparation of the poultry for sale. In this process, the bird is killed by stunning, followed by bleeding, scalding, defeathering, washing, evisceration, and finally chilling. Gunderson *et al.* (1954), Clark and Lentz (1969a), Surkiewicz *et al.* (1969), and others have studied the effect of these processes on the total levels of bacteria in poultry. Although levels of bacteria on the skin do not significantly change from the live bird to the final chilled carcass (10^3 to 6 × 10^4/cm^2), some increase usually results from the evisceration process. Further washing and chilling, however, reduces the contamination, but there is not a single processing step which drastically reduces the bacterial count. The psychrophilic count also gradually increases throughout the process to levels of about 10^3/cm^2. The microflora after processing therefore consists of a wide range of organisms including *Acinetobacter, Moraxella, Cytophaga, Flavobacterium, Pseudomonas, Enterobacter, Chromobacterium, Aeromonas, Proteus, Escherichia, Salmonella, Alcaligenes, Campylobacter, Bacillus, Clostridium*, and *Micrococcus*.

The surface microflora which eventually dominates will depend on the holding temperature, pH, and the presence or absence of O_2 and carbon dioxide. At refrigeration temperatures (about 1 °C) and in the presence of O_2, the pseudomonads usually dominate. Organisms such as *Pseudomonas putida, P. fluorescens*, and *P. putrefaciens* attain levels of 10^9/cm^2, giving rise to off odors and slime formation (Ayres *et al.*, 1950; Nagel and Simpson, 1960). At higher temperatures (15 °C), enterobacteria and lactobacilli may become the major groups (Barnes and Shrimpton, 1968).

Work by Barnes and Impey (1968) demonstrated that the difference in pH of the chicken breast and leg tissues could lead to variation in the genera that were able to grow. During the early stages of growth at 1 °C in minced breast muscle, *Acinetobacter* species were the dominant group. After about 4 days, the nonpigmented pseudomonads became the major group, and after 10 days, they were the only group present. In minced leg muscle at 1 °C, the *Acinetobacter* species were able to dominate up to about 4 days, but again the nonpigmented and the pigmented pseudomonads became the major group of organisms. Unlike the breast muscle experiments, however, the *Acinetobacter* were still in evidence after 6–7 days and formed 20% of the population.

If poultry carcasses are packaged in an O_2-impermeable film, the growth of the pseudomonads is controlled. Under these conditions, the lactobacilli, *M. thermosphactum*, and enterobacteria become the major organisms present on the surface of the poultry (Barnes and Shrimpton, 1968; Arafa and Chen, 1975).

Although the major part of the microbial activity is on the surface of the carcasses, penetration of bacteria into the underlying tissues can take place. This occurs mainly via exposed inner surfaces of the body cavity, e.g., cut surfaces at the neck flap, vent, and broken skin. Pseudomonads are located chiefly around these surfaces, and they are usually not found in the deep underlying muscle tissue (Sanders, 1969; Patterson, 1972; Barnes 1975).

3.4. Red Meats

Red meats as a habitat for microorganisms have been studied extensively (Ayres, 1955; Kitchell and Ingram, 1956; Ingram, 1962; Pierson *et al.*, 1970; Ingram and Dainty, 1971; Surkiewicz *et al.*, 1975; Ingram and Roberts, 1976; Sunderland *et al.*, 1977). Meat has a high moisture content (a_w >0.96), a pH within the range of 5.5–7.0, a low redox potential, and high protein-to-carbohydrate ratio (19:1). Thus, it is a medium which supports a range of pathogenic and nonpathogenic microorganisms. Healthy animals usually have little bacterial contamination of their tissues (Gill *et al.*, 1976). The flora of meat is therefore due to invaders. The practices employed in slaughter or during chilling, dressing, and cutting of the carcasses affect the ease with which meat becomes invaded. The bacteria come mainly from the animal's hide, respiratory tract, or gastrointestinal tract. Thus, the meat, after chilling and cutting, contains a far wider range of organisms, mostly consisting of saprophytic species, than the original tissue. Ayres (1951) lists numerous genera on fresh meats, including *Micrococcus, Staphylococcus, Pseudomonas, Flavobacterium, Streptococcus, Escherichia, Bacillus, Microbacterium*, and several genera of yeasts and molds.

One of the principal factors which determines the dominance of the various species listed is storage temperature. Most fresh meats are held at temperatures below 10°C to improve the shelf life, causing the preferential growth of psychrophiles such as pseudomonads, which develop during storage. Other groups of organisms which may not be present at slaughter but are important after chill storage are *Microbacterium, Acinetobacter, Moraxella*, and LAB. Growth is thought to start at the surface of the meat. Figure 6 is an illustration of how the growth of the major groups of organisms could take place under aerobic conditions. An interesting species in the community is *M. thermosphactum*. This organism is unique to the meat environment and has not been reported in any other food. Its source is unknown. The properties of *M. thermosphactum* were first reported by McLean and Sulzbacher (1953). They showed that lactic acid is the end product of its energy metabolism, and that it has low heat resistance and is facultatively anaerobic. Its ability to grow was further explored under

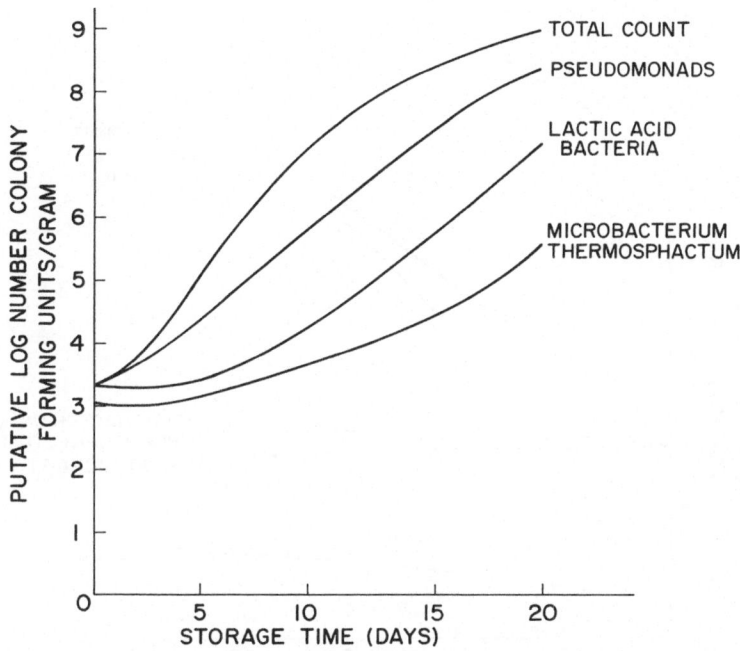

Figure 6. Probable development of the microflora of red meat during refrigerated storage in air.

aerobic conditions by Brownlie (1966) and under anaerobic conditions by Weidemann (1965). The organism thrives under reduced O_2 supply and increasing carbon dioxide concentration (Sunderland *et al.*, 1977). Carbon dioxide is less inhibitory to *M. thermosphactum* than to other aerobic organisms, e.g., the pseudomonads. Gardener and Carson (1967) demonstrated that low concentrations of carbon dioxide stimulated its growth. This is not surprising because this organism fixes carbon dioxide for the synthesis of aspartate (Collins-Thompson *et al.*, 1970). The similarity between *M. thermosphactum* and certain LAB was also discussed in Section 2.4.1 (see Fig. 2). *M. thermosphactum* and LAB thrive in the same ecosystem.

The various LAB associated with fresh meat tissues include *Lactobacillus viridans, L. brevis,* and *L. plantarum* (Reuter, 1975). Under aerobic conditions and at temperatures of less than 10°C, these organisms can grow, but, nevertheless, the pseudomonads are the dominant group. If, however, O_2 is limited at the surface of the meat, a shift in dominance is observed (Fig. 7). The pseudomonads can no longer compete successfully, and the LAB become the dominant group. The behavior of *M. thermosphactum* is unpredictable under these conditions, and the literature contains contradictory results. Pierson *et al.* (1970) reported a

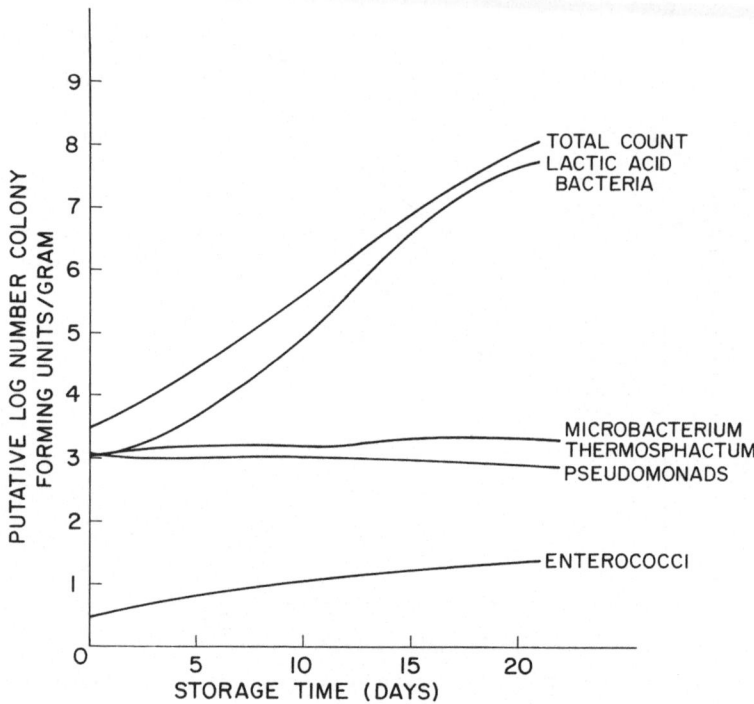

Figure 7. Probable development of the microflora of red meat during refrigerated vacuum-packed storage.

decrease in the population of *M. thermosphactum* in vacuum-packaged beef stored at $-3\,°C$, while Sunderland *et al.* (1977) reported an initial increase. The reason for this difference is not known, but it may well be related to the controlling effect of the LAB. Roth and Clarke (1975) suggest that the presence of sufficient numbers of lactobacilli can inhibit the multiplication of *M. thermosphactum*.

The multiplication of *Pseudomonas* or other gram-negative bacteria is limited by O_2 and the accumulated carbon dioxide in the package. Levels of 20% carbon dioxide have been shown to be most effective in inhibiting these organisms (Sunderland *et al.*, 1977).

A special mention should also be made of *Leuconostoc lactis*. This organism has been found in large numbers in vacuum-packaged beef (Pierson *et al.*, 1970), where it accounted for the difference between the count of LAB and total count after storage for 15 days at $-3\,°C$. A comparison between Figs. 6 and 7 shows that there is a difference in the total count reached. The maximum number in aerobically packaged meats exceeds that in anaerobically packaged meat. This fact was noted by Ingram (1962) and has never been fully explained. Ingram suggested that certain metabolic disturbances due to pH changes and O_2 limitation may play an important role in limiting total numbers. The breakdown of meat tissues under aerobic conditions is more extensive than under anaerobic conditions (Sunderland *et al.*, 1976), and additional nutrients released during aerobic breakdown might support further growth.

The role of pH is important not only in limiting the total population in meat tissues, but also in selecting the genera supported by meat. Inoculation studies by Patterson and Gibbs (1977) using meat of normal pH (5.4) and meat of high pH (6.6) indicated that the two types of meat did not support the same groups of organisms. With meat of high pH value, after 8 weeks of storage at 0-2°C in vacuum packs, psychrotrophic enterobacteria represented a large proportion of the microflora ($10^6/cm^2$). Although such organisms have been reported in stored fresh meats (Gardener and Carson, 1967; Ingram and Dainty, 1971), they do not appear to play a major role in the spoilage of meat of normal pH value. Enterococci have been shown to be present in normal meat but do not appear to be major spoilage organisms (Pierson *et al.*, 1970).

The microbiology of meat changes as the storage temperature is increased to 10°C. There is then a shift from the psychrophilic population to mesophilic organisms such as *Bacillus* and *Micrococcus* (Ayres, 1951).

The microbiological breakdown of meats has been considered a surface phenomenon, and work by Dainty *et al.* (1975) and Gill and Penney (1977) has indicated that this is partially true. When certain proteolytic surface bacteria such as the pseudomonads reach their maximum cell density in meats, they secrete extracellular proteases. These enzymes break down the connective meat tissue between the muscle fibers, allowing such bacteria to penetrate the meat. Unless the meat is in an advanced stage of decomposition, however, the rate of penetration is limited.

3.5. Cured Meats

3.5.1. Bacon

The microbiology of sliced bacon in vacuum packs was reviewed by Hurst (1973). The normal flora of the product spoiling at room temperature consists mostly of micrococci and LAB, which cause a "souring" type of spoilage. Pathogenic staphylococci do not develop at 20°C or below. High numbers of staphylococci do develop at 30°C, but enterotoxin tends not to be formed.

Inoculation of bacon stored at above 20°C caused putrefaction, but at temperatures below 20°C, fecal streptococci tended to dominate, spoiling the bacon by souring. In many Western countries, nisin is a permitted food additive, but its use in vacuum-packed sliced bacon did not delay spoilage (Gibbs and Hurst, 1964).

3.5.2. Ham

Ham made from raw meat is a valuable product and is widely manufactured in America and Europe. Microbial interactions in French and Italian hams were reviewed by Hurst (1973). Leistner and Rödel (1976b) summarize the effects of manufacturing processes in inhibiting *C. botulinum* and *Trichinella spiralis*, a parasite found in pork muscle. They stress the importance of low water activity; the parasites are no longer invasive when the a_w is decreased to <0.93.

The microbiology of raw ham is quite different from that of canned ham. The former keeps indefinitely at room temperature because of low a_w, whereas canned ham, which receives a relatively mild heat treatment but has a high a_w content, requires refrigerated storage for good stability. Enterococci may survive the heat treatment, and they can antagonize surviving bacilli (Hurst, 1973). *C. perfringens* spores which survive the heating process may be prevented from outgrowth by ingredients of curing salts (Labbe and Duncan, 1970).

3.5.3. Sausages

Sausages are made of fermented or unfermented meat, each of which may or may not be cooked. Figure 8, which shows this relationship, greatly underestimates the complexity of the available products. In the first instance, each of the four categories can be further subdivided according to whether or not the products are smoked. Over 100 varieties of the heated "emulsion" products of the bologna- and frankfurter-type sausages can be recognized in Europe (Leistner

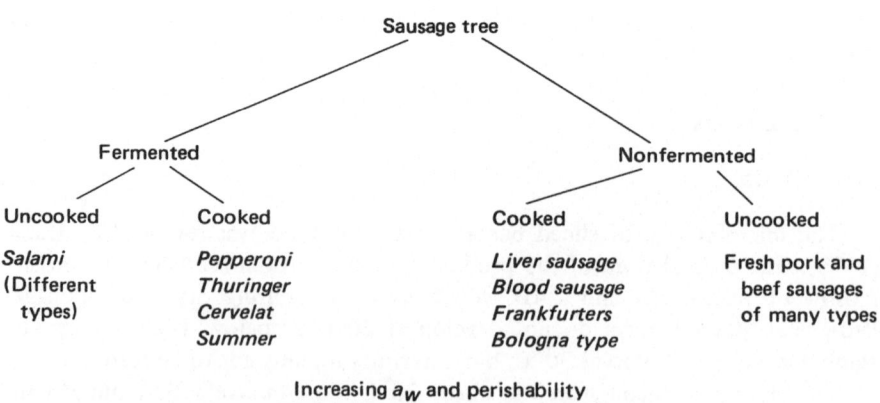

Figure 8. A sausage tree, to illustrate the microbiological relatedness of different kinds of sausage. Most products are made from cured comminuted meat, usually seasoned, usually stuffed into casings. The resulting sausage products may or may not be smoked, according to different manufacturers and differing national habits.

and Rödel, 1976b). There are types of sausage which cannot be fitted into any scheme, for example, a blood sausage (which needs cooking) containing pieces of fat and pieces of cured meat and/or tongue. The composition of these foods will evidently influence the microbial flora which develops.

The most stable products are on the left of Fig. 8. These are salamis made from fermented raw meat; these sausages have low a_w values and contain salt and fat. These factors add up to products which have long keeping quality and require no refrigerated storage. As we move toward the right of Fig. 8, we encounter products of increasing perishability, chiefly because of increasing pH and a_w values.

The fermentation of meat is comparable to that of milk, and its microbiology was extensively studied at the American Meat Institute Foundation in the early 1960s (Deibel *et al.*, 1961). Most present-day authorities recommend the use of starters consisting of cultures of LAB, e.g., *P. cerevisiae*, lactic streptococci, or antibiotic-producing micrococci. The rationale for the use of the nonlactic organisms is that a strain of micrococcus, M_{53}, produces an antibiotic inhibitory to many of the undesirable bacteria present in the raw sausage mix. LAB appear to be resistant to the M_{53} antibiotic so that they develop, reducing the E_h and pH. In conjunction with smoking, these factors are sufficient to kill the antibiotic-producing micrococci (Hurst, 1973). Starter cultures are not used by all companies manufacturing fermented sausages. In such cases, fresh but fully fermented sausage is added back to the meat mix to act as a source of LAB. This practice has the advantage of economy but the disadvantage that, in general, there is no knowledge about the nature and the properties of the flora being added.

In modern practice, two further steps are recommended which ensure the rapid dominance of the added starter culture: first, addition of glucose (about 1%) to the ground-meat mixture because, as explained below, the shortage of carbohydrate in meat discriminates against a lactic fermentation. Second, even a small quantity of a chemical acidulant helps to ensure the dominance of LAB (Genigeorgis, 1976). Glucono delta lactone (GDL) and citric acid are frequently used; the former compound hydrolyzes in meat to release gluconic acid, which gently lowers meat pH in a manner resembling that occuring in a fermentation. The pH reached depends on the amount of GDL used.

The major microbial change that occurs during natural fermentation of meat is a shift from the flora of fresh refrigerated meat (gram negative, catalase positive, aerobic) to a lactic acid flora (gram positive, catalase negative, microaerophilic). The combined efforts of the LAB and man result in the lowering of the pH, a_w, and redox potential and the formation of inhibitors, e.g., antibiotics and hydrogen peroxide. The necessary storage temperature of these products and their keeping quality will then depend on the extent of these interacting biochemical transformations.

The next question concerns the safety of the products (for a review, see Genigeorgis, 1976). *S. aureus* especially causes concern because of its association with the skin of animals (and certain cheap cuts, which are popular in sausage manufacture). Thus, it may be introduced into the sausage mix. In addition, none of the individual factors listed above (pH, a_w, E_h, and inhibitors) can be relied on singly to inhibit staphylococci. The relatively good record of fermented sausages relies on the simultaneous application of all the limiting physicochemical factors together with appropriate cooling. Some guidelines and prediction of shelf life are offered by the work of Leistner and Rödel (1976a). Salmonellae also are often found in fresh meat, but salmonellosis is only rarely attributable to the consumption of fermented meats. In properly prepared fermented meat, the degree of their survival depends on the weight of the initial contamination. The use of starter reduces the chances of their survival (Genigeorgis, 1976).

3.6. Dairy Products

The microbiology of margarine, butter, and certain shelf-stable dairy products (condensed, evaporated, and dried milk) will not be discussed. Each of these products has a large dairy literature related to its technology, keeping quality, and safety.

3.6.1. Milk

Dairy products start with milk which, contrary to common belief, is seldom sterile. Carefully drawn milk generally contains microbes picked up from the udder cistern and teat canal. If the cow is healthy, these organisms should be harmless, but pathogenic organisms are known to be secreted in milk.

"Milk in a container" is a highly perishable and delicate product. It is readily contaminated by the flora from the animal's skin, the air (which can be an indirect fecal contamination), utensils, wash water, etc. Pathogenic bacteria can be introduced from these sources. To increase safety and keeping quality, milk is almost invariably heated, most commonly by the high-temperature short-time (HTST) (about 73 °C for 15 sec) pasteurization treatment. Ultrahigh-temperature (UHT) milk (about 125 °C for 5 sec) is also being introduced gradually, especially in Europe. All these treatments destroy most of the non-spore-forming flora so that pasteurized milk left at room temperature does not clot and sour. Indeed, it is the writers' experience that *S. lactis* may be difficult to isolate, even from raw milk. Possibly this is due to the complete disappearance of wooden dairy utensils and the high efficiency of the detergents and sanitizers in use at present.

Fecal streptococci and *Microbacterium* may, partially at least, survive HTST. *Bacillus* and *Clostridium* spores also survive but are generally prevented from developing if the milk is stored at about 5°C. Storage at 10°C or higher permits many spore formers to develop. However, in modern practice, it is the psychrotrophic flora which are particularly important (e.g., pseudomonads, which are a postprocessing contamination). They are capable of developing from very small inocula to high numbers (e.g., 10^6/ml) within a few days of refrigerated storage. These organisms are destroyed by HTST, but their enzymes are not. Dairy products made from such milk may develop off-flavors due to heat-resistant proteolytic and lipolytic enzymes originally made by the psychrotrophs. This defect is difficult to diagnose. It is aggravated by modern practices such as bulk collection systems and the nonworking of dairies during weekends.

3.6.2. Fermented Products

3.6.2a. General. The fermentation of lactose by the LAB, with the concomitant fall in pH, formation of peroxides, antibiotics, etc. (see Section 2.4) is a remarkably efficient way of preserving dairy products. Other animal products, e.g., meat, fish, and eggs, do not "sour" in the way milk does. These deteriorate more by putrefaction than by souring, first, because of the higher fermentable sugar content of milk and, second, because of its homogeneity, which ensures the absence of pockets of decay. Davis (1975) comments on how fortunate it is that nature has put lactose into milk rather than some other sugar. Lactose always undergoes a lactic fermentation with acid production, which is a powerful mechanism for the prevention of the development of pathogens. For example, *E. coli* is never found in yogurt after the first day of storage (Davis, 1975).

Fermented dairy products can be broadly divided into two categories: first, those which are consumed fresh, e.g., yogurt or lactic cheese. In these products, lactose is hydrolyzed, but the protein and lipid components are unchanged. Products in the second category undergo a "ripening" period after the lactic fermentation when the protein and lipid portions also undergo profound biochemical transformations. Hard cheeses like Cheddar, Emmental, and Gouda are examples of this second category; they may be ripened for 6 months or more, during which time important changes occur in the bacterial flora. Between these extremes, there exists a wide range of intermediate products.

3.6.2b. Yogurt. After a rather severe heating, the milk is cooled to 42°- 45°C and inoculated with a starter culture of *S. thermophilus* and *L. bulgaricus.* The milk should clot in about 3 hr. After the attainment of the desired degree of acidity, the product is stored in the refrigerator. Provided the container is solid and the air space is small, the product is almost indefinitely stable—a constrast to the milk from which it was made. The end of organoleptic acceptability (generally about 20 days) is reached for nonbacteriological reasons, i.e., excessive whey separation or excessive acidity which continues to develop even at refrigeration temperatures.

The streptococcus and the lactobacillus coexist synergistically in this product. The mixed culture forms acid more rapidly than either culture alone because each produces compounds stimulatory to the growth of the other culture. Therefore, if one or the other member of this finely balanced mixed culture is damaged, acid production slows down. This happens when *S. thermophilus* suffers a bacteriophage attack, which also slows down the growth of the lactobacillus. Slow acid production results in economic loss and the possibility that some food-poisoning organisms could develop, e.g., *S. aureus*.

Flavored and fruit yogurts are popular items, but they present problems. Sugar and/or fruit juices must be added before heating to secure adequate pasteurization. If the heating is diminished to avoid caramelized flavor, yeasts may survive. Yeasts are not inhibited by the LAB and grow with them. They may utilize the lactate, or they may yield alcohol from lactose, as is the case in some deliberate mixed fermentations resulting in kefir or koumis.

It is probable that in these products, symbiotic relationships exist between the LAB and the yeasts, though this does not appear to have been studied extensively, and the basis of the interaction is little understood. The "grains" which develop in kefir are themselves evidence of an intimate interaction. It may be mentioned in passing that the leaven of the bread referred to in the Old Testament was probably a mutually beneficial association of yeasts and LAB similar to that discussed here. This ferment is still used for the production of sourdough bread (Wood *et al.*, 1975).

3.6.2c. Cheese. Cheese and other fermented milk foods are the subject of a 700-page book by Kosikowski (1977). In this chapter, it is only possible to sketch in bare outlines.

Processed cheese is made from natural cheese heated with emulsifiers (e.g., polyphosphates) and various ingredients. Spore-forming organisms can survive this treatment, and a lactate-utilizing species, *Clostridium tyrobutyricum*, can form gas and spoil processed cheese made from Swiss-type cheese but not English cheese. Fortunately, the presence of this organism in processed cheese can be controlled. Spores are introduced with the natural cheese to which they gained access from milk. In its turn, milk is most commonly infected by silage being fed to cows. The habitat of this organism thus appears to be grass; the feeding of silage to cows in countries where Swiss-type cheese is produced is prohibited by law.

Figure 9 is an attempt to display the microbiological relatedness of different kinds of cheese. In all cheese making, the major protein of milk (casein) is precipitated, and this precipitate is used to trap the butter fat. At the top of the illustration, the precipitation is shown to be carried out by acid, since casein precipitates at its isoelectric point. When the acid formed by microbes hydrolyz-

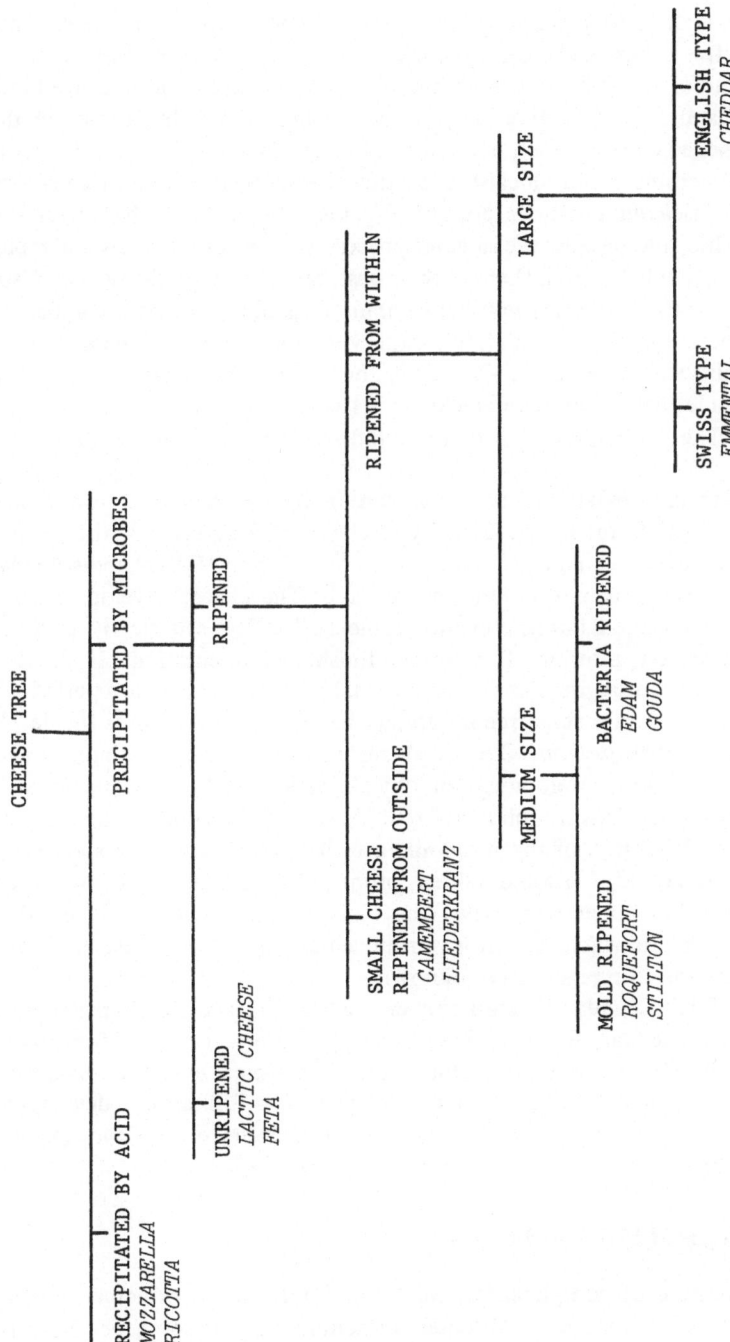

Figure 9. A cheese tree, to illustrate the microbiological relatedness of different kinds of cheese.

ing lactose is used to precipitate the curd, cottage cheese is obtained. This is consumed fresh. Other cheeses are ripened to varying degrees. With the exception of some of the Italian cheeses, the curd is obtained by the combined effect of acidity and a milk-clotting enzyme, e.g., rennin. Basically, there are three kinds of ripened cheese: small, medium, and large. The small ones can be ripened by strictly aerobic molds localized at the cheese surface (e.g., Camembert). *Penicillium caseicolum* is the responsible organism. Such cheeses have a relatively short shelf life, and because the ripening process goes on continuously and rapidly, it is often difficult to catch them at their best. Medium-sized cheese may also be mold ripened; this mold, *Penicillium roqueforti*, is blue, so that the appearance of this cheese is quite different from the white-mold-ripened cheese. This organism is also strictly aerobic. Therefore, the cheese cannot be too large. It has to be so made that it develops cracks or it has to be punctured to permit O_2 to penetrate. The O_2 requirement of the mold is what determines the size of the cheese.

The large cheeses are ripened from within by an anaerobic or microaerophilic lactic acid flora. In the Swiss-type cheese-making process, the curds are heated to about $55°C$, and the large cheese mass (up to 100 kg) cools slowly to permit the development of thermophilic LAB. The cheese develops a skin as the result of repeated immersion into brine baths. The salt slowly penetrates toward the cheese interior. The correct finishing fermentation pH should be about 5.2, and the pH rises as the cheese matures. Thus, there is a period when the interior of the cheese is free of salt and has a pH which permits the development of *C. tyrobutyricum*. Clostridial gas formation leads to large irregular pockets formed and development of objectionable taste; in extreme cases, the cheese may burst. This is quite different from the desirable gas formation resulting from the action of the propionic acid bacteria. These are responsible for the "eye formation." Carbon dioxide is produced from lactate a few weeks after the cheese-making operation, when the body of the cheese is still elastic. In Cheddar cheese, in which the salt is introduced as part of the cheese-making process, similar faults are practically unknown.

"Germ-free" Cheddar cheese has been made but lacked the characteristics of conventional Cheddar, showing that bacterial action is necessary for obtaining cheese flavor. The starter culture alone gave some aroma, but other organisms in addition, especially lactobacilli, are necessary for full aroma development. Enzymes present in raw milk also probably contribute to full flavor (Law and Sharpe, 1975).

3.7. Catalogue of Fermented Foods

For the sake of completeness, we list in Table III the principal fermented foods used by Western man. Oriental fermentations and soy sauce have been omitted. Some of the items in Table III were discussed in Sections 3.5 and 3.6;

Table III. Partial List of Fermented Foods[a]

Product	Original flora and its fate	Starter microflora or dominant flora, at end of process
1. Fermented milk, cream, butter, English-type cheese	Original microflora is destroyed by heating; survivors are a few enterococci and spore formers.	Starter of mesophilic LAB which dominate.
2. Yogurt	As no. 1.	Thermophilic LAB which dominate.
3. Kefir		Inoculated with kefir grains, a physical association of yeasts and LAB.
4. Soft cheese (small)	As no. 1.	As no. 1, plus *Penicillium caseicolum*; some types of small soft cheese are surface ripened by *Brevibacterium linens*.
5. Blue cheese	As no. 1.	As no. 1, plus spores of *Penicillium roqueforti*.
6. Swiss-type cheese	As no. 1.	As no. 2; propionic acid bacteria added for "eye" formation.
7. Fermented meats (sausages)	Organisms from animal hide: micrococci, staphylococci; organisms from feces: enterobacteria, enterococci, clostridia; organisms from water: pseudomonads, proteus; organisms from air: spore formers; in pasteurized meat products heat-resistant flora might partially survive, e.g., some enterococci and some spore formers.	*Pediococcus cerevisiae* and antibiotic-producing micrococci added, and the former survives; mesophilic LAB dominate, including lactobacilli; the surface of some salami contains lipolytic yeasts to improve appearance and flavor.
8. Fermented vegetables (sauerkraut, olives)	Enterobacteria and LAB originally present.	Starter not usually necessary to ensure dominance of LAB.
9. Bread	Aerobic and anaerobic spore formers; yeasts.	Similar to no. 3.

[a]Abbreviation: LAB, lactic acid bacteria.

others, such as sauerkraut, have not been discussed but are mentioned in this table.

4. Protection and Injury of Bacteria by Food-Processing Treatments

4.1. General

Bacteria used to be considered dead when they were unable to form visible colonies on "good" media. It is now accepted that growth on a medium may not reflect the bacteria contained in the sample because various treatments and bacterial interactions can "injure" microbes. Injury is manifested by a long lag phase and an inability to grow on selective media (van Schothorst, 1976; Busta, 1976; Hurst, 1977).

Modern studies reveal conditions which not only injure but also protect bacteria.

4.2. Protection

Lowered water activity increases the heat resistance of *Salmonella*, but there is no direct correlation between a_w and heat resistance. Sucrose especially is far more effective at comparable water activities in protecting salmonellae than other solutes (Baird-Parker *et al.*, 1970; Goepfert *et al.*, 1970). Corry (1976) suggested that the degree of protection can be correlated with the degree of plasmolysis, i.e., that heat resistance is due to cell dehydration. Other studies have also shown that ions can be protective; for example, Mg^{2+} was protective against heat damage in *Aerobacter aerogenes* (Strange and Shon, 1964) and in *S. typhimurium* (Lee and Goepfert, 1975).

Fat can also be highly protective (Senhaji, 1977). The lipids of chocolate may explain the survival and pathogenicity of *Salmonella eastbourne* in this product. In the course of manufacture, beans were roasted and ground to a liquor which was held at 65°C for 48 hr (D'Aoust *et al.*, 1975). The material was further held for 12–24 hr at 60°C. It is difficult to imagine how salmonellae could survive this treatment. Bacteria can be protected against damage due to freezing by cryoprotectants. This subject is complicated and has been reviewed recently by MacLeod and Calcott (1976).

4.3. Injury

The practical significance of injured microbes in food processing and technology is concerned with the maintenance of starter activity and uniformity of fermentations. Modern techniques have eliminated starter handling in the cheese factory and the lag phase associated with cultures handled by mail. Briefly, these techniques involve growing the LAB with pH control to avoid excessive acidity during growth, concentrating the culture by centrifuging, and preserving with liquid N_2.

In food preservation, the practical interest in injured microbes is exactly opposite to that of the considerations listed for starters. The aim is to minimize processing and to ensure that the surviving bacteria have extremely long lag phases and are inhibited by reagents which may not affect uninjured microbes. For example, Roberts and Ingram (1966) showed that spores surviving heat treatments were considerably more sensitive to the curing salts commonly used in meats than were unheated spores.

Recovering spores damaged by heat, radiations, or ethylene oxide was extensively discussed by Roberts (1970). Spore injury may become evident at any of three stages of development: the germination system, the formation of the first delicate vegetative cell, and outgrowth. Of these, practically the most important one is damage to the germination system. Cassier and Sebald (1968) were first to show that heat may destroy an enzyme system required for spore germination in *C. perfringens* without otherwise affecting viability. The germination of these spores could be restored by lysozyme added to the medium. Later, Sebald and Ionesco (1972) showed that similar considerations apply to *C. botulinum* type E, which was previously thought to be temperature sensitive. Instead, it appears that the only heat-labile part of the spores is an enzyme system concerned with germination, and viability could be restored by adding lysozyme to the medium.

5. Concluding Remarks

In this chapter, we considered foods which deteriorate rather than canned or other foods of great shelf stability. The process of deterioration can result in food spoilage or development of food-borne hazards. The process starts from the moment of harvesting (or milk production or slaughter, etc.) with the initial microflora or the first contamination. The composition of this initial microflora changes continuously due to physical, chemical, and biological factors and their interactions.*

Three principal categories of these factors were recognized by Mossel and Ingram (1955), namely (a) intrinsic factors which are the subject of the discus-

---

*In general, bacteria place little demand on their environment and can be expected to grow rapidly. Microbial activity ceases when: (a) special procedures are used to inhibit bacteria (freezing) or to kill them (heating); (b) a combination of limiting factors is used, e.g., low pH plus low a_w plus low storage temperature; and (c) in fermented foods, lactate accumulates; such foods frequently harbor pure cultures of LAB, and lactate is the end product of their metabolism. Cases (a) to (c) represent acceptable foods. Exhaustion of nutrients [case (d)] could also be expected to stop microbial activity, but such food would likely be unacceptable.

sion below, (b) extrinsic factors (water activity, temperature of storage, and partial pressure of O_2), which will not be discussed further, and (c) implicit factors, i.e., the nature of the microbial flora of a food. This was discussed in the previous sections, and also see Mossel (1971).

The intrinsic factors influencing the development of food microflora are physical factors (Section 5.1), chemical factors (Section 5.2), and interactions (Section 5.3).

5.1. Physical Factors

The rate of development of a food microflora depends in part on structural factors, e.g., whether the food is a homogeneous liquid or a solid containing areas which differ from one another. An example of this was quoted in Section 3.3: The dark and white poultry meat generally have different pH values. Other physical factors concern emulsions. For example, margarine and butter are water-in-oil emulsions with bacteria living in the aqueous phase. Bacterial development depends on droplet size; small droplets might contain insufficient nutrients to support a spoilage population. Other examples are intact biological barriers which hinder microbial growth, e.g., the exoskeleton of crustaceans or the membrane of eggs.

5.2. Chemical Factors

Food deterioration starts at the expense of the most readily available and soluble components, e.g., carbohydrates, peptides, amino acids, etc. (Ingram and Dainty, 1971). Protein breakdown of aerobically stored meat does not start until an advanced stage of spoilage (Dainty et al., 1975). Lipolytic enzymes become important only in the later stages of deterioration. The amount of fats present in food is often comparable to that of other ingredients; the later utilization of fats may be connected with their relative insolubility.

The literature contains many reports of the antibacterial properties of some foods, especially some spices and vegetables (e.g., onion, garlic, cinnamon, etc.) (Zaika et al., 1978). Similarly, colostrum and raw milk are known to possess antibacterial properties (Reiter, 1976). However, what is more surprising is that Zaika et al. (1978) report that the lactic fermentation, which is required for the preparation of Lebanon-bologna-type sausage, only occurs normally in the presence of spices. Without spices, the salt content of the meat is sufficient to delay the fermentation.

5.3. Interactions

Antibiotic production in foods by different groups of microbes was reviewed by Hurst (1973). Interactions between different physical factors have also been frequently studied. Interaction of cold with many physical factors has been re-

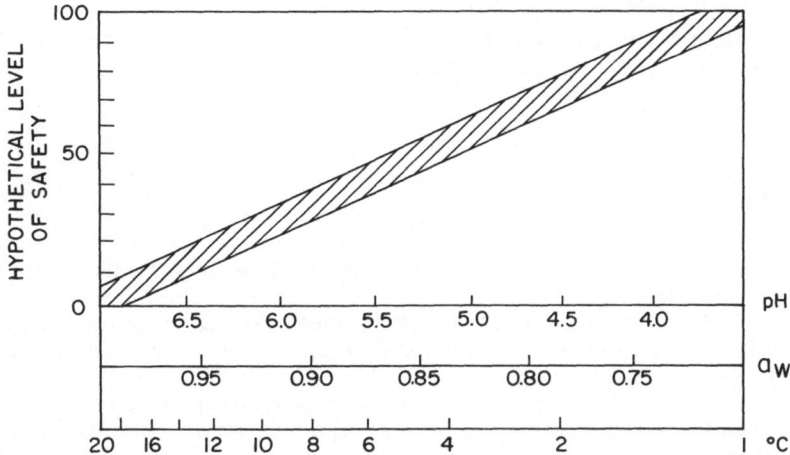

Figure 10. Hypothetical diagram to illustrate the possible correlation between physicochemical parameters of foods and their safety.

viewed by Ingram and Mackey (1976). Their figure 3 shows growth of *S. faecalis* var. *liquefaciens* affected by a four-factor interaction among pH, sodium chloride content, water activity, and temperature. At 20–35°C, the organism grows at any pH from 5.7 to 6.6 at water activities of 0.925–0.990 and concentrations of sodium chloride of up to 12.3%. At 10°C, the pH range remains unchanged, but the minimum a_w increases from 0.925 to 0.945, and the maximum level of sodium chloride diminishes from 12.3% to 9.6%. Diminishing the incubation temperature even further to 5°C narrows the pH limits of growth to 6.0–6.6, the minimum a_w increases to 0.965, and the maximum concentration of sodium chloride tolerated is 6.5%. Incubation at 1°C stops all growth. Nitrite, an important preservative in cured meats (see Section 2.5), similarly shows interactions between pH and sodium chloride content (Pivnick and Barnett, 1965). In many foods, such interactions are now known to occur but were first developed empirically. This is discussed by Leistner and Rödel (1976a). Some of the interactions listed by them are, for example, cooking temperature, pH, and reduced water activity, which serve to preserve jams; other interactions are used to preserve meats, cakes, dried fruits, and frozen foods.

An illustration of hypothetical interactions between some of the physicochemical parameters of food was suggested by H. Sinell (Fig. 10). This figure is used as an illustration of interactions and should not be used to determine safety to foods in practice. It shows that a food at pH 3.5 or below is likely to be safe.* A food at pH 5.0, however, is unlikely to keep well on account of a

---

*Reservations are necessary to this statement, which is true of a food uncontaminated with salmonellae or some other pathogens. These are likely to survive storage at this pH, especially at low temperatures. A pH value of 3.5 also does not guarantee freedom from mold growth and the possibility of mycotoxin synthesis.

pH effect alone. We are looking for another "50%" of safety action. This could be supplied by a storage temperature of 4°C or by a storage temperature of 8°C and a reduction of water activity to about 0.90.

References

Alexander, M., 1971, Amensalism, in: *Microbial Ecology*, pp. 301-302, John Wiley and Sons, New York.

Alford, J. A., Pierce, D. A., and Sulzbacher, W. L., 1964, Microbial lipases and their potential importance to the meat industry, *Proc. 15th Res. Conf. Res. Counc. Am. Meat Inst. Found. Univ. Chicago*, No. 74, pp. 11-16.

Alm, F., Ericksen, I., and Molin, N., 1961, The effect of vacuum packaging on some sliced processed meat products as judged by organoleptic and bacteriological analysis, *Food Technol.* 15:199-203.

Anonymous, 1969, An Evaluation of the *Salmonella* Problem, Report, National Academy of Sciences, Washington, D.C.

Arafa, A. S., and Chen, T. C., 1975, Effect of vacuum packaging on microorganisms on cut up chickens and in chicken products, *J. Food Sci.* 40:50-53.

Ashworth, J., and Spencer, R., 1972, The Perigo effect in pork, *J. Food Technol.* 1:111-124.

Ayres, J. C., 1951, Some bacteriological aspects of spoilage of self service meats, *Iowa State Coll. J. Sci.* 26:31-48.

Ayres, J. C., 1955, Microbiological implications in handling, slaughtering and dressing of meat animals, *Adv. Food Res.* 6:109-161.

Ayres, J. C., Ogilvy, W. S., and Steward, G. F., 1950, Post mortem changes in stored meats. I. Microorganisms associated with the development of slime on eviscerated cut up poultry, *Food Technol.* 4:199-205.

Baird-Parker, A. C., Boothroyd, M., and Jones, E., 1970, The effect of water activity on the heat resistance of heat sensitive and heat resistant strains of salmonellae, *J. Appl. Bacteriol.* 33:515-522.

Bala, K., Marshall, R. T., Stringer, W. C., and Naumann, H. D., 1977, Effect of *Pseudomonas fragi* on the color of beef, *J. Food Sci.* 42:1176-1179.

Barnes, E. M., 1975, The microbiological problems of sampling a poultry carcass, in: *Proceedings of the 2nd European Symposium on Poultry Meats*, Oosterbeck, The Netherlands.

Barnes, E. M., and Impey, C. S., 1968, Psychrophilic spoilage bacteria of poultry, *J. Appl. Bacteriol.* 31: 97-107.

Barnes, E. M., and Shrimpton, D. H., 1968, The effects of processing and marketing procedures on the bacteriological condition and shelf life of eviscerated turkeys, *Br. Poult. Sci.* 9:243-251.

Barrow, G. I., 1963, The nature of inhibitory action by *Staphylococcus aureus* Type 71, *J. Gen. Microbiol.* 32:255-261.

Bartl, V., 1973, Semi-preserved foods: General microbiology and food poisoning, in: *The Microbiological Safety of Food* (B. C. Hobbs and J. H. B. Christian, eds.), pp. 89-106, Academic Press, London.

Bennett, R., Taylor, D. R., and Hurst, A., 1966, D- and L-lactate dehydrogenases in *Escherichia coli*, *Biochim. Biophys. Acta* 118:512-524.

Bergdoll, M. S., 1970, Enterotoxins, in: *Microbial Toxins* (T. C., Montie, S. Kadie, and S. J. Ajl, eds.), Vol. III, pp. 265-326, Academic Press, New York.

Board, R. G., 1970, Microbiology of the hen's egg, *Adv. Appl. Microbiol.* 11:245–251.

Board, R. G., and Fuller, R., 1974, Non-specific antimicrobial defences of the avian egg, embryo and neonate, *Biol. Rev.* 49:15–21.

Briston, J., 1976, New plastic materials for packaging films, *Packag. Rev.* 71–74.

Brownlie, L. E., 1966, Effects of some environmental factors on psychrophilic microbacteria, *J. Appl. Bacteriol.* 29:447–454.

Buchanan, R. E., and Gibbons, N. E. (eds.), 1974, *Bergey's Manual of Determinative Bacteriology*, Williams and Wilkins, Baltimore.

Busta, F. F., 1976, Practical implications of injured microorganisms in food, *J. Milk Food Technol.* 39:138–145.

Cassier, M., and Sebald, M., 1968, Germination lysozyme-dépendante des spores des *Clostridium perfringens* ATCC 3624 après traitement thermique, *Ann. Inst. Pasteur Paris* 117:312–324.

Cavett, J. J., 1962, The microbiology of vacuum packed sliced bacon, *J. Appl. Bacteriol.* 25:282–289.

Cavett, J. J., 1968, The effects of newer forms of packaging on the microbiology and storage life of meats, poultry and fish, in: *Progress in Industrial Microbiology* (D. J. D. Hockenhull, ed.), pp. 77–123, J. and A. Churchill, London.

Clark, D. S., and Lentz, C. P., 1969a, Microbiological studies in poultry processing plants in Canada, *Can. Inst. Food Sci. Technol. J.* 2:33–36.

Clark, D. S., and Lentz, C. P., 1969b, The effect of carbon dioxide on the growth of slime producing bacteria on fresh beef, *Can. Inst. Food Sci. Technol. J.* 2:72–75.

Collins-Thompson, D. L., Witter, L. D., and Ordal, Z. J., 1970, Carbon dioxide fixation and synthesis of aspartate by *Microbacterium thermosphactum*, *Biochem. Biophys. Res. Commun.* 40:909–913.

Collins-Thompson, D. L., Sørhaug, T., Witter, L. D., and Ordal, Z. J., 1972, Taxonomic consideration of *Microbacterium lacticum, Microbacterium flavum* and *Microbacterium thermosphactum*, *Int. J. Syst. Bacteriol.* 22:65–72.

Collins-Thompson, D. L., Aris, B., and Hurst, A., 1973, Growth and enterotoxin B synthesis by *Staphylococcus aureus* S6 in associative growth with *Pseudomonas aeruginosa*, *Can. J. Microbiol.* 19:1197–1201.

Colwell, R. R., 1962, The bacterial flora of Puget Sound fish, *J. Appl. Bacteriol.* 25:147–158.

Cords, B. R., McKay, L. L., and Guerry, P., 1974, Extrachromosomal elements in the group of streptococci, *J. Bacteriol.* 117:1149–1152.

Corry, J. E., 1976, Sugar and polyol permeability of *Salmonella* and osmophilic yeast cell membranes measured by turbidimetry, and its relation to heat resistance, *J. Appl. Bacteriol.* 40:277–284.

Coulthard, C. E., Michaelis, R., Short, W. F., Sykes, G., Skrimshire, G. E. H., Standfast, A. F. B., Birkinshaw, J. H., and Raistrick, H., 1945, Notatin: An anti-bacterial glucose-aerodehydrogenase from *Penicillium notatum* Westling and *Penicillum resticulosum* sp. nov., *Biochem. J.* 39:24–36.

Cramer, J. A., and Prestegard, J. H., 1977, NMR studies of pH-induced transport of carboxylic acids across phospholipid vesicle membranes, *Biochem. Biophys. Res. Commun.* 75:205–301.

Dainty, R. H., Shaw, B. G., deBoer, K. A., and Sheps, E. S. J., 1975, Protein changes caused by bacterial growth on beef, *J. Appl. Bacteriol.* 39:73–81.

D'Aoust, J. Y., Aris, B. J., Thisdele, P., Durante, A., Brisson, N., Dragou, D., Lachapelle, G., Johnston, M. and Laidley, R., 1975, *Salmonella eastbourne* outbreak associated with chocolate, *Can. Inst. Food Sci. Technol. J.* 8:181–184.

Davis, J. G., 1975, The microbiology of yoghurt, in: *Lactic Acid Bacteria in Beverages and Food* (J. G. Carr, C. V. Cutting, and G. C. Whiting, eds.), pp. 254–263, Academic Press, London.

Defigueiredo, M. P., and Splittstoesser, D. F., 1976, *Food Microbiology: Public Health and Spoilage Aspects*, Avi Publishing Co., Westport, Connecticut.

Deibel, R. H., Niven, C. F., and Wilson, D. D., 1961, Microbiology of meat curing. III. Some microbiological and related technological aspects in the manufacture of fermented sausages, *Appl. Microbiol.* 9:156–161.

Dolman, C. E., 1964, Botulism as a world health problem, in: *Botulism* (K. H. Lewis and K. Cassel, eds.), p. 5, U.S. Department of Health, Education, and Welfare, Public Health Service, Cincinnati, Ohio.

Dornbusch, K., and Hallander, H. D., 1973, Transduction of penicillinase production and methicillin resistance to enterotoxin B production in strains of *Staphylococcus aureus*, *J. Gen. Microbiol.* 76:1–11.

Dowdell, M. J., and Board, R. G., 1968, A microbiological survey of British fresh sausage, *J. Appl. Bacteriol.* 31:378–396.

Eddy, B. P., and Kitchell, A. G., 1959, Cold tolerant fermentative gram-negative organism from meat and other sources, *J. Appl. Bacteriol.* 22:57–61.

Elliot, R. P., and Michener, D. H., 1965, Factors Affecting the Growth of Psychrophilic Micro-organisms in Foods. A Review, Technical Bulletin No. 1320, Agricultural Research Service, U.S.D.A.

Empey, W. A., and Scott, W. J., 1939, Investigations on chilled beef. I. Microbial contamination acquired in the meatworks, *Aust. Commonw. Counc. Sci. Ind. Res. Bull.* 126:1–71.

Felton, E. A., Evans, J. B., and Niven, C. F., Jr., 1953, Production of catalase by pediococci, *J. Bacteriol.* 65:481–482.

Frazier, W. C., 1968, *Food Microbiology*, McGraw-Hill, New York.

Gale, E. F., 1962, The development of a good round character, *J. Appl. Bacteriol.* 25:309–323.

Gardener, G. A., and Carson, A. W., 1967, Relationship between carbon dioxide production and growth of pure strains of bacteria on porcine muscle, *J. Appl. Bacteriol.* 30:500–510.

Genigeorgis, C. A., 1976, Quality control of fermented meats, *J. Am. Vet. Med. Assoc.* 169:1220–1228.

Ghuysen, J. M., 1968, Use of bacteriolytic enzymes in determining of wall structure and their role in cell metabolism, *Bacteriol. Rev.* 32:425–464.

Gibbs, B. M., and Hurst, A., 1964, Limitations of nisin as a preservative in non-dairy foods, in: *Microbial Inhibitors in Food* (N. Molin, ed.), pp. 151–165, Almqvist and Wiksell, Stockholm.

Gill, C. O., and Penney, N., 1977. Penetration of bacteria into meat, *Appl. Environ. Microbiol.* 33:1284–1286.

Gill, C. O., and Suisted, J. R., 1978, The effects of temperature and growth rate on the proportion of unsaturated fatty acid in bacterial lipids, *J. Gen. Microbiol.* 104:31–36.

Gill, C. O., Penney, N., and Nottingham, P. M., 1976, Effect of delayed evisceration on the microbial quality of meat, *Appl. Environ. Microbiol.* 31:456–468.

Gillespie, N. C., and Macrae, I. C., 1975, The bacterial flora of some Queensland fish and its ability to cause spoilage, *J. Appl. Bacteriol.* 39:91–100.

Gilliland, S. E., and Speck, M. L., 1975, Inhibition of psychrotrophic bacteria by lactobacilli and pediococci in non-fermented refrigerated foods, *J. Food Sci.* 40:903–905.

Goepfert, J. M., and Kim, H. U., 1975, Behaviour of selected food-borne pathogens in raw ground beef, *J. Milk Food Technol.* 38:449–452.

Goepfert, J. M., Iskander, I. K., and Amundson, C. H., 1970, Relation of heat resistance of salmonellae to water activity of the environment, *Appl. Microbiol.* 19:429–433.

Goepfert, J. M., Spiro, W. M., and Kim, H. U., 1972, *Bacillus cereus:* Food poisoning organism. A review, *J. Milk Food Technol.* 35:213–227.

Gould, G. W., 1969, Germination, in: *The Bacterial Spore* (G. W. Gould and A. Hurst, eds.), pp. 397–444, Academic Press, London.

Gunderson, M. F., McFadden, H. W., and Kyle, T. S., 1954, *The Bacteriology of Commercial Poultry Processing*, Burgess Publishing Co., Minneapolis, Minnesota.

Haines, R. B., 1938, Observations on the bacterial flora of the hen's egg, with a description of new species of *Proteus* and *Pseudomonas* causing rots in eggs, *J. Hyg.* 38:338–341.

Haines, W. C., and Harmon, L. G., 1973, Effect of selected lactic acid bacteria on growth of *Staphylococcus aureus* and production of enterotoxins, *Appl. Microbiol.* 25:436–441.

Harrison, J. M., and Lee, J. S., 1969, Microbial evaluation of Pacific shrimp processing, *Appl. Microbiol.* 18:188–192.

Harry, E. G., 1963, The relationship between egg spoilage and the environment of the egg when laid, *Br. Poult. Sci.* 4:91–94.

Hauge, S., 1955, Food poisoning caused by aerobic sporeforming bacilli, *J. Appl. Bacteriol.* 18:591–595.

Hauschild, A. H. W., 1974, Food poisoning by *Clostridium perfringens*, in: *Anaerobic Bacteria: Role in Disease* (A. Balows, R. M. DeHaan, V. R. Dowell, and L. B. Guze, eds.), pp. 149–155, C. C. Thomas, Springfield, Illinois.

Hirsch, A., 1952, The evolution of the lactic streptococci, *J. Dairy Res.* 19:290–293.

Hirsch, A., McClintock, M., and Mocquot, G., 1952, Observations on the influence of inhibitory substances produced by the lactobacilli of Gruyère cheese on the development of anaerobic spore-formers, *J. Dairy Res.* 19:179–186.

Hobbs, G., 1976, *Clostridium botulinum* and its importance in fishery products, *Adv. Food Res.* 22:135–185.

Hsu, C. Y., and Wiseman, G. M., 1971, Purification of epidermidins, new antibiotics from staphylococci, *Can. J. Microbiol.* 17:1223–1226.

Hurst, A., 1973, Microbial antagonism in foods, *Can. Inst. Food Sci. Technol. J.* 6:80–90.

Hurst, A., 1977, Bacterial injury: A review, *Can. J. Microbiol.* 23:935–944.

Hurst, A., 1978, Nisin: Its preservative effect and function in the growth-cycle of the producer organism, in: *The Streptococci* (F. A. Skinner and J. Quesnell, eds.), pp. 297–314, Academic Press, London.

Hurst, A., and Kruse, H., 1972, Effect of secondary metabolites on the organisms producing them: Effect of nisin on *Streptococcus lactis* and enterotoxin B on *Staphylococcus aureus*, *Antimicrob. Agents Chemother.* 1:277–279.

Hurst, A., Hughes, A., and Collins-Thompson, D. L., 1974, Effect of sublethal heating on *Staphylococcus aureus* at different physiological ages, *Can. J. Microbiol.* 20:765–768.

Hurst, A., Hendry, G. S., Hughes, A., and Paley, B., 1976, Enumeration of sublethally heated staphylococci in some dried foods, *Can. J. Microbiol.* 22:677–683.

Iandolo, J. J., Clark, C. W., Blum, L., and Ordal, Z. J., 1965, Repression of *Staphylococcus aureus* in associative culture, *Appl. Microbiol.* 13:646–649.

Ingram, M., 1962, Microbiological principles in prepacking meats, *J. Appl. Bacteriol.* 25:259–281.

Ingram, M., 1969, Sporeformers as food spoilage organisms, in: *The Bacterial Spore* (G. W. Gould and A. Hurst, eds.), pp. 549–610, Academic Press, London.

Ingram, M., 1975, The lactic acid bacteria—a broad view, in: *Lactic Acid Bacteria in Beverages and Food* (J. G. Carr, C. V. Cutting, and G. C. Whiting, eds.), pp. 1–13, Academic Press, London.

Ingram, M., and Dainty, R. H., 1971, Changes caused by microbes in spoilage of meats, *J. Appl. Bacteriol.* 34:21–39.

Ingram, M., and Mackey, B. M., 1976, Inactivation by cold, in: *Inhibition and Inactivation of Vegetative Microbes* (F. A. Skinner and W. B. Hugo, eds.), pp. 111–151, Academic Press, London.

Ingram, M., and Roberts, T. A., 1976, The microbiology of red meat carcass and the slaughter house, *R. Soc. Health J.* 96:270–276.

Ingram, M., Ottaway, F. J. H., and Coppock, J. B. M., 1956, The preservative action of acid substances in food, *Chem. Ind. (London) Part 2*, pp. 1154–1162.

Ivanovics, G., 1962, Bacteriocins and bacteriocin-like substances, *Bacteriol. Rev.* 26:108–118.

Johannsen, A., 1965, *Clostridium botulinum* type E in foods and the environment generally, *J. Appl. Bacteriol.* 28:90–94.

Kao, C. T., and Frazier, W. C., 1966, Effect of lactic acid bacteria on growth of *Staphylococcus aureus*, *Appl. Microbiol.* 14:251–255.

Kato, N., and Shibasaki, I., 1975, Comparison of antimicrobial activities of free fatty acids and their esters, *J. Ferment. Technol.* 53:793–801.

Katz, E., and Demain, A. L., 1977, The peptide antibiotics of *Bacillus*: Chemistry, biogenesis and possible functions, *Bacteriol. Rev.* 41:449–474.

Kempton, A. G., and Bobier, S. R., 1970, Bacterial growth in refrigerated vacuum packed luncheon meats, *Can. J. Microbiol.* 16:287–297.

Kim, H. U., and Goepfert, J. M., 1971, Occurrence of *Bacillus cereus* in selected dry food products, *J. Milk Food Techol.* 34:12–15.

Kitchell, A. G., 1962, Micrococci and coagulase negative staphylococci in cured meats and meat products, *J. Appl. Bacteriol.* 25:416–431.

Kitchell, A. G., and Ingram, M., 1956, A comparison of bacterial growth on fresh and on frozen meat after thawing, *Ann. Inst. Pasteur Lille* 8:121–131.

Kitchell, A. G., and Shaw, B. G., 1975, Lactic acid bacteria in fresh and cured meat, in: *Lactic Acid Bacteria in Berverages and Food* (J. G. Carr, C. V. Cutting, and G. C. Whiting, eds.), pp. 209–220, Academic Press, London.

Kosikowski, F., 1977, *Cheese and Fermented Milk Foods*, Edwards Brothers, Ann Arbor, Michigan.

Labbe, R. G., and Ducan, C. L., 1970, Growth from spores of *Clostridium perfringens* in the presence of sodium nitrite, *Appl. Microbiol.* 19:353–359.

Law, B. A., and Sharpe, M. E., 1975, Lactic acid bacteria and cheese flavour, in: *Lactic Acid Bacteria in Beverages and Food* (J. G. Carr, C. V. Cutting, and G. C. Whiting, eds), pp. 233–244, Academic Press, London.

Laycock, R. A., and Loring, D. H., 1972, Distribution of *Clostridium botulinum* type E in the Gulf of St. Lawrence in relation to the physical environment, *Can. J. Microbiol.* 18:763–773.

Laycock, R. A., and Regier, L. W., 1970, Pseudomonads and Actinobacter in the spoilage of irradiated haddock of different preirradiation quality, *Appl. Microbiol.* 20:333–341.

Lee, A. C., and Goepfert, J. M., 1975, Influence of selected solutes on thermally induced death and injury of *Salmonella typhimurium*, *J. Milk Food Technol.* 38:195–200.

Lee, C. N., and Foster, E. M., 1965, Effect of vacuum packaging on growth of *Clostridium botulinum* and *Staphylococcus aureus* in cured meats, *Appl. Microbiol.* 13:1023–1025.

Lee, W. H., 1977, An assessment of *Yersinia enterocolitica* and its presence in foods, *J. Food Prot.* 40:486–489.

Lee, W. H., Staples, C. L., and Olson, J. C., Jr., 1975, *Staphylococcus aureus* growth and survival in macaroni dough and the persistence of enterotoxins in the dried products, *J. Food Sci.* 40:119–120.

Leistner, L., and Rödel, W., 1976a, The stability of intermediate moisture foods with respect to micro-organisms, in: *Intermediate Moisture Foods* (R. Davies, G. G. Birch, and K. J. Parker, eds.), pp. 120–134, Applied Science Publishers, London.

Leistner, L., and Rödel, W., 1976b, Inhibition of micro-organisms in food by water activity, in: *Inhibition and Inactivation of Vegetative Microbes* (F. A. Skinner and W. B. Hugo, eds.), pp. 219–237, Academic Press, London.

Lerke, P., Adams, R., and Farber, L., 1963, Bacteriology of spoilage of fish muscle. I. Sterile press juice as a suitable experimental medium, *Appl. Microbiol.* 11:458–462.

Lerke, P., Adams, R., and Farber, L., 1965, Bacteriology of spoilage of fish muscle. III. Characteristics of spoilers, *Appl. Microbiol.* 13:625–630.

Lifshitz, A. R., Baker, C., and Naylor, H. B., 1964, The relative importance of chicken egg exterior structures in resisting bacterial penetration, *J. Food Sci.* 29:94–99.

Lifshitz, A. R., Baker, C., and Naylor, H. B., 1965, The exterior structures of the egg as nutrients for bacteria, *J. Food Sci.* 30:516–519.

MacLeod, R. A., and Calcott, P. H., 1976, Cold shock and freezing damage to microbes, *Symp. Soc. Gen. Microbiol.* 26:81–109.

Marples, M. J., 1965, *The Ecology of the Human Skin*, C. C. Thomas, Springfield, Illinois.

Mattick, A. T. R., and Hirsch, A., 1944, Sour milk and tuberculosis, *Lancet* 1:417.

McCoy, D. W., and Faber, J. E., 1966, Influence of food microorganisms on staphylococcal growth and enterotoxin production, *Appl. Microbiol.* 14:372–377.

McLean, R. A., and Sulzbacher, W. L., 1953, *Microbacterium thermosphactum* spec. nov: A non heat resistant bacterium from fresh pork sausage, *J. Bacteriol.* 65:428–433.

Meers, J. L., 1973. Growth of bacteria in mixed culture, *CRC Crit. Rev. Microbiol.* 2:139–184.

Meyer, K. F., 1956, The status of botulism as a world health problem, *Bull. W. H. O.* 15:281.

Miller, A., Scanlan, R. A., Lee, J. S., and Libbey, L. M., 1973, Identification of the volatile compounds produced in sterile fish muscle (*Sebastes melanops*) by *Pseudomonas fragi, Appl. Microbiol.* 25:952–955.

Mocquot, G., 1971, Occurrence and role of microorganisms in cheese, in: *International Symposium on Conversion and Manufacture on Foodstuffs*, Kyoto, Japan, pp. 191–197.

Morse, S. A., and Mah, R. A., 1973, Regulation of staphylococcal enterotoxin B: Effect of anaerobic shock, *Appl. Microbiol.* 25:553–557.

Mossel, D. A. A., 1971, Physiological and metabolic attributes of microbial groups associated with foods, *J. Appl. Bacteriol.* 34:95–118.

Mossel, D. A. A., and Ingram, M., 1955, The physiology of the microbial spoilage of foods, *J. Appl. Bacteriol.* 18:232–268.

Nagel, C., and Simpson, K., 1960, Microorganisms associated with spoilage of refrigerated poultry, *Food Technol.* 14:21–23.

Nickerson, J. N., and Sinskey, A. J., 1974, *Microbiology of Foods and Food Processing*, p. 74, American Elsevier, New York.

Niinvaara, F. P., 1955, Über den Einfluss von Bakterienkulturen auf die Reifung und umrotung der Rohwurst, *Acta Agric. Fenn.* 84:1–128.

Niinivaara, F. P., and Pohja, M. S., 1957, Erfahrungen bei der Herstellung von Rohwurst mit Bakterienkulturen, *Fleischwirtschaft* 9:789–790.

Niskanen, A., and Nurmi, E., 1976, Effect of starter culture on staphylococcal enterotoxin and thermonuclease production in dry sausage, *Appl. Environ. Microbiol.* 31:11–20.

Oberhofer, T. R., and Frazier, W. C., 1961, Competition of *Staphylococcus aureus* with other organisms, *J. Milk Food Technol.* 24:172–175.

Oblinger, J. L., and Kraft, A. A., 1970, Inhibitory effects of *Pseudomonas* on selected *Salmonella* and bacteria isolated from poultry, *J. Food Sci.* 35:30–33.

Oblinger, J. L., and Kraft, A. A., 1973, Oxidation and reduction potential and growth of *Salmonella* and *Pseudomonas flourescens, J. Food Sci.* 38:1108–1112.

Orla-Jensen, A., 1919, *The Lactic Acid Bacteria*, Andr. Fred. Host and Son, Copenhagen.

Panes, J. J., and Thomas, S. B., 1968, Psychrotrophic coli-aerogenes bacteria in refrigerated milk: A review, *J. Appl. Bacteriol.* 31:420–425.

Patterson, J. T., 1972, Microbiological sampling of poultry carcasses, *J. Appl. Bacteriol.* 35: 569–575.

Patterson, J. T., and Gibbs, P. A., 1977, Incidence and spoilage potential of isolates from vacuum-packaged meat of high pH value, *J. Appl. Bacteriol.* 43:25–38.

Pederson, C. S., and Ward, L., 1949, The effect of salt upon the bacteriological and chemical changes in fermenting cucumbers. *Tech. Bull. N.Y.S. (Geneva) Agric. Exp. Sta.* 288: 1–29.

Pelroy, G. A., Seman, J. P., Jr., and Eklund, M. W., 1967, Changes in the microflora of irradiated petrale sole (*Eopsetta jordani*) fillets stored aerobically at 0.5°C, *Appl. Microbiol.* 15:92–96.

Pierson, M. D., Collins-Thompson, D. L., and Ordal, Z. J., 1970, Microbiological, sensory and pigment changes of aerobically and anaerobically packaged beef, *Food Technol.* 24:129–133.

Pivnick, H., and Barnett, H., 1965, Effect of salt and temperature on toxigenesis by *Clostridium botulinum* in perishable cooked meats vacuum packed in air-impermeable plastic pouches, *Food Technol.* 19:1164–1167.

Pivnick, H., and Chang, P. C., 1973, Perigo effect in pork, in: *Proceedings of the International Symposium on Nitrite in Meat Products* (B. Krol and B. J. Tinbergen, eds.), pp. 111–116, Pudoc, Wageningen, The Netherlands.

Rayman, M. K., Park, C. E., Philpott, J., and Todd, E. D., 1975, Reassessment of the coagulase and thermostable nuclease tests as a means of identifying *Staphylococcus aureus, Appl. Microbiol.* 29:451–454.

Reddy, M. S., Vedamuthu, E. R., Washam, C. J., and Reinbold, G. W., 1971, Associative growth relationships in two strain mixtures of *Steptococcus lactis* and *Streptococcus cremoris, J. Milk Food Technol.* 34:236–240.

Reddy, S. G., Henrickson, R. L., and Olson, H. C., 1970, The influence of lactic cultures on ground beef quality, *J. Food Sci.* 35:787–791.

Reiter, B., 1976, Bacterial inhibitors in milk, in: *Inhibition and Inactivation of Vegetative Microbes* (F. A. Skinner and W. B. Hugo, eds.), pp. 31–60, Academic Press, London.

Reuter, G., 1971, Laktobazillen und eng verwandte Mikroorganismen in Fleisch and Fleischwaren, *Fleischwirtschaft* 51:1237–1242.

Reuter, G., 1975, Classification problems, ecology and some biochemical activities of lactobacilli of meat products, in: *Lactic Acid Bacteria in Beverages and Food* (J. G. Carr, C. V. Cutting, and G. C. Whiting, eds.), pp. 221–229, Academic Press, London.

Riemann, H., 1973, Botulinum food poisoning, *Can. Inst. Food Sci. Technol. J.* 6:111–125.

Roberts, T. A., 1970, Recovering spores damaged by heat, ionizing radiations or ethylene oxide, *J. Appl. Bacteriol.* 33:74–94.

Roberts, T. A., and Ingram, M., 1966, The effect of sodium chloride, potassium nitrate and sodium nitrite on the recovery of heated bacterial spores, *J. Food Technol.* 1:147–163.

Rose, A. H., 1962, Temperature relationships among microorganisms, *Wallerstein Lab. Commun.* 25(86):5–18.

Roth, L. A., and Clarke, D. S., 1975, Effect of lactobacilli and carbon dioxide on the growth of *Microbacterium thermosphactum* on fresh beef, *Can. J. Microbiol.* 21:629–632.

Sanders, D. H., 1969, Fluorescent dye tracing of water-entry and retention in chilling of broiler chicken carcasses, *Poult. Sci.* 48:2032-2037.

Schiemann, D. A., 1976, Occurrence of *Klebsiella pneumoniae* in dairy products, *J. Milk Food Technol.* 39:467-469.

Sebald, M., and Ionesco, H., 1972, Germination lz-dépendante des spores de *Clostridium botulinum* type E, *C. R. Acad. Sci.* 275D:2175-2177.

Sen, N. P., 1974, Nitrosamines, in: *Toxic Constituents of Animal Foodstuffs* (I. E. Liener, ed.), pp. 131-158, Academic Press, New York.

Senhaji, A. F., 1977, The protective effect of fat on the heat resistance of bacteria (II), *J. Food Technol.* 12:217-230.

Shalita, Z., Hartman, I., and Sarid, S., 1977, Isolation and characterization of a plasmid involved with enterotoxin B production in *Staphyloccus aureus, J. Bacterol.* 129: 317-325.

Shaw, B. G., and Shewan, J. M., 1968, Psychrophilic spoilage bacteria in fish, *J. Appl. Bacteriol.* 31:89-92.

Shelef, L. A., 1977, Effect of glucose on the bacterial spoilage of beef, *J. Food Sci.* 42: 1172-1175.

Shewan, J. M., 1961, The microbiology of sea water fish, in: *Fish as Food* (G. Borstrom, ed.), pp. 487-560, Academic Press, London.

Shewan, J. M., 1962, The bacteriology of fish and spoiling fish and some related chemical changes, *Recent Adv. Food Sci.* 1:167-193.

Shewan, J. M., 1971, The microbiology of fish and fishery products—a progress report, *J. Appl. Bacteriol.* 34:299-315.

Shewan, J. M., and Hobbs, G., 1967, The bacteriology of fish spoilage and preservation, *Prog. Ind. Microbiol.* 6:171-179.

Shewan, J. M., and Jones, N. R., 1957, Chemical changes occurring in cod muscle during chill storage and their possible use as objective indices of quality, *J. Sci. Food Agric.* 8:491-498.

Snow, J. E., and Beard, P. J., 1939, Studies on the bacterial flora of North Pacific salmon, *Food Res.* 4:563-585.

Stamer, J. R., 1975, Recent developments in the fermentation of sauerkraut, in: *Lactic Acid Bacteria in Beverages and Food* (J. G. Carr, C. V. Cutting, and G. C. Whiting, eds.), pp. 267-280, Academic Press, London.

Strange, R. E., and Shon, M., 1964, Effects of thermal stress on viability and ribonucleic acid of *Aerobacter aerogenes* in aqueous suspensions, *J. Gen. Microbiol.* 34:99-114.

Strasters, K. C., and Winkler, K. C., 1963, Carbohydrate metabolism of *Staphylococcus aureus, J. Gen. Microbiol.* 33:213-229.

Su, T. L., 1948, Micrococcin. An antibacterial substance formed by a strain of micrococcus, *Br. J. Exp. Pathol.* 29:473-481.

Sugiyama, H., and Yang, K. H., 1975, Growth potential of *Clostridium botulinum* in fresh mushrooms packaged in semi-permeable plastic film, *Appl. Microbiol.* 30:964-969.

Sunderland, J. P., Gibbs, P. A., Patterson, J. T., and Murrey, J. G., 1976, Biochemical changes in vacuum packaged beef occurring during storage at 0.2°C, *J. Food Technol.* 11:171-180.

Sunderland, J. P., Patterson, J. T., Gibbs, P. A., and Murrey, J. G., 1977, The effect of several gaseous environments on the multiplication of organisms isolated from vacuum-packaged beef, *J. Food Technol.* 12:249-256.

Surkiewicz, B. F., Johnston, R. W., Moran, A. B., and Krumm, G. W., 1969, A bacteriological survey of chicken eviscerating plants, *Food Technol.* 23:80-85.

Surkiewicz, B. F., Harris, M. E., Elliot, R. P., Macaluso, J. F., and Strand, M. M., 1975, Bacteriological survey of raw beef patties produced at establishments under federal inspection, *Appl. Microbiol.* 29:331–334.

Tadd, A. D., and Hurst, A., 1961, The effect of feeding colicinogenic *Escherichia coli* on the intestinal *E. coli* of early weaned pigs, *J. Appl. Bacteriol.* 24:222–228.

Thimann, K. V., 1963, *The Life of the Bacteria*, Macmillan, New York.

Troller, J. A., and Christian, J. H. B., 1978, *Water Activity and Food*, Academic Press, London.

Troller, J. A., and Frazier, W. C., 1963, Repression of *Staphylococcus aureus* by food bacteria. II. Causes of inhibition, *Appl. Microbiol.* 11:163–165.

van Schothorst, M., 1976, Resuscitation of injured bacteria in foods, in: *Inhibition and Inactivation of Vegetative Microbes* (F. A. Skinner and W. B. Hugo, eds.), pp. 317–328, Academic Press, London.

Vanderzant, C., Mroz, E., and Nickelson, J. N., 1970, Microbial flora of Gulf of Mexico and pond shrimp, *J. Milk Food Technol.* 33:346–350.

Vasconcelos, G. J., and Lee, J. S., 1972, Microbial flora of Pacific oysters (*Crassostrea gigas*) subject to ultraviolet-irradiated seawater, *Appl. Microbiol.* 23:11–16.

Vincent, J. G., Veomett, R. C., and Riley, R. F., 1959, Antibacterial activity associated with *Lactobacillus acidophilus*, *J. Bacteriol.* 78:477–484.

Walsh, D. E., and Funke, B. R., 1975, The influence of spaghetti extruding, drying and storage on survival of *Staphylococcus aureus*, *J. Food Sci.* 40:714–716.

Watson, D. W., 1939, Studies of fish spoilage. V. The role of trimethylamine oxide in the respiration of *Achromobacter*, *J. Fish. Res. Board Can.* 4:267–280.

Weidemann, J. R., 1965, A note on the microflora of beef stored in nitrogen at 0°C, *J. Appl. Bacteriol.* 28:365–367.

Wheater, D. M., Hirsch, A., and Mattick, A. T. R., 1952, Possible identity of "Lactobacillin" with hydrogen peroxide produced by lactobacilli, *Nature (London)* 170:623–624.

Whiteside, T. L., and Corpe, W. A., 1969, Extracellular enzymes produced by a *Pseudomonas* sp. and their effect on cell envelopes of *Chromobacterium violaceum*, *Can. J. Microbiol.* 15:81–92.

Whittenbury, R., 1978, Biochemical characteristics of *Streptococcus* spp.: 1) Aerobic properties; 2) are Pediococci and Aerococci Streptococci? in: *The Streptococci* (F. A. Skinner and J. Quesnel, eds.), pp. 51–69, Academic Press, London.

Williams, J. E., Dillard, L. H., and Hall, G. O., 1968, Penetration patterns of *Salmonella typhimurium* through the outer structures of chicken eggs, *Avian Dis.* 12:445–466.

Witter, L. D., 1961, Psychrophilic bacteria, A review, *J. Dairy Sci.* 44:983–1015.

Wood, B. J., Cardemas, O. S., Yong, F. M., and McNulty, D. W., 1975, Lactobacilli in production of soy sauce, sour-dough bread and Parisian Barm, in: *Lactic Acid Bacteria in Beverages and Food* (J. G. Carr, C. V. Cutting, and G. C. Whiting, eds.), pp. 325–335, Academic Press, London.

Zaika, L. L., Zell, T. E., Palumbo, S. A., and Smith, J. L., 1978, Effect of spices and salt on fermentation of Lebanon bologna-type sausage, *J. Food Sci.* 43:186–189.

Zyskind, J. W., Pattee, P. A., and Lache, M., 1965, Staphylolytic substance from a species of *Pseudomonas*, *Science* 147:1458–1459.

4

Microbial Ecology of the Oral Cavity

G. H. W. BOWDEN, D. C. ELLWOOD, AND I. R. HAMILTON

1. Introduction

The importance of the oral microorganisms to the process of dental caries was first recognized by Miller (1889) at the end of the last century. In his chemico-parasitic theory, he stated that carbohydrate food particles were decomposed to organic acids by bacteria, which resulted in the decalcification of the enamel surface of the tooth. Despite this early recognition that the mouth carried an indigenous microflora, it is only in the last 18 years that any interest has been shown in oral microbial ecology. Earlier studies had concentrated on the role of specific acidogenic bacteria in dental caries and often neglected the possibility of the influence of microbial interactions on this and other oral diseases, such as periodontal disease. One of the earlier workers to recognize such interactions was MacDonald in his studies of the pathogenicity of oral organisms in oral infections. He and his co-workers, in a variety of studies on mixed anaerobic infections (MacDonald *et al.*, 1954, 1960, 1963), employed combinations of pure cultures of oral anaerobic organisms to demonstrate that minimum synergistic groupings were capable of producing pathogenic lesions in guinea pigs.

G. H. W. BOWDEN • MRC Dental Epidemiology Unit, London Hospital Dental School, London Hospital Dental School, London, England. Present address: Department of Oral Biology, Faculty of Dentistry, University of Manitoba, Winnipeg, Canada. D. C. ELL-WOOD • Microbiological Research Establishment, Porton Down, England. I. R. HAMILTON • Department of Oral Biology, Faculty of Dentistry, University of Manitoba, Winnipeg, Canada.

These lesions possessed the same inflammatory and suppurative characteristics as those observed in advanced periodontal lesions.

The influence of medical microbiological concepts (i.e., Koch's postulates) led many workers to search for a single etiological agent for dental caries and periodontal disease. Only now has this approach led, finally, to the concept that in both of these common oral diseases the ecology of the microbial communities is a major factor.

2. Bacteria in the Oral Environment

2.1. Introduction

In common with other areas of the human intestinal tract, the oral cavity supports a community of commensal bacteria. Several reviews cover various aspects of this microbial community (Hardie and Bowden, 1974a, 1976a; Gibbons and van Houte, 1975a,b; Stiles *et al.*, 1976).

The oral flora develops in complexity up to adult life (McCarthy *et al.*, 1965; Socransky and Manganiello, 1971; Long and Svenson, 1976; Hardie and Bowden, 1976a). A relatively simple flora exists in newborn children, with *Streptococcus salivarius* being a predominant organism. It is possible that contaminating organisms from the vagina and rectum of the mother do not persist in the newborn's mouth since fecal lactobacilli do not become established (Carlsson and Gothefors, 1975).

Attempts to establish bacteria in mouths carrying a commensal flora have proved difficult (Krasse *et al.*, 1967; Mikx *et al.*, 1976b; Svanberg and Loesche, 1978a,b; van der Hoeven and Rogers, 1978) even when oral species are used. Implanted *Streptococcus* strains usually only persist for short periods and fail to become established. Transmission of indigenous organisms from habitat to habitat within the same mouth does seem to be a possibility (Mikx and Svanberg, 1978), although populations are often localized in a specific habitat (van Houte, 1976). Intrafamilial transfer of oral bacteria has been explored in the case of *Streptotoccus mutans*, but again there is little evidence that this occurs easily (Melisch *et al.*, 1978). However, there does seem to be a relationship between the number of *S. mutans* in the saliva of mothers and colonization in their children (Kohler and Bratthall, 1976). Certainly the numbers of specific organisms in saliva can influence their colonization of oral surfaces (van Houte, 1976; Duchin and van Houte, 1978). It seems likely that organisms from the external environment, even if they are of oral origin, cannot readily establish themselves in a mouth carrying a normal commensal flora. The reasons for this are not well known, but it must be assumed that several factors play a role, perhaps the most significant being other organisms occupying all available niches in the oral ecosystem.

The bacteria isolated from areas within the mouth have developed characteristics which enable them to survive against the removal effects of saliva, its antibacterial activity, immunological mechanisms, and the competition of other members of the community. As a result organisms can be isolated which are unique to the mouth, while others are oral representatives of genera present as commensals in other areas of the body (Gibbons et al., 1963, 1964; Handleman and Mills, 1965; Gordon and Gibbons, 1966; Gordon and Jong, 1968; Hadi and Russell, 1968; Bowden and Hardie, 1971, 1978; Socransky and Manganiello, 1971; Loesche et al., 1972; Loesche and Syed, 1973; Edwardsson, 1974; Hardie and Bowden, 1974a, 1976a; Bowden et al., 1975; Socransky, 1977).

The mouth provides a series of habitats for colonization by bacteria, and it is unique in that the tooth is a nonshedding surface, unlike the mucosa of other areas. A wide variety of environmental conditions results from the interrelationships of tissue surfaces, saliva, and enamel. Moreover, in most instances, the nutritional status of the mouth is complex, and normally an intermittent supply of varied and relatively rich nutrients is available to the flora. The diversity of the environments provided in the mouth is reflected in the complexity of the flora, which ranges from aerobic nonsaccharolytic organisms to sensitive anaerobes requiring specific nutrients, growth factors, and conditions (Hardie and Bowden, 1974a; Gibbons and van Houte, 1975a; Socransky, 1977).

In early studies on the bacteriology of the mouth, the environment was often considered as a whole, and the bacteria in saliva were often taken as representative of the flora of the specific habitats. In some cases, this type of examination was used to assess the extent or potential amount of carious lesions in a subject (Ellen, 1976; Bowden et al., 1978).

More recently, examinations have been made on more localized communities in well-defined areas in the mouth, particularly on the surfaces of the teeth. The results of these studies have demonstrated the specific nature of the microbial communities and have also served to focus attention on the ecology of various habitats in the mouth (Loesche and Syed, 1973; Theilade et al., 1973; Edwardsson, 1974; Hardie and Bowden, 1974a; Bowden et al., 1975; Gibbons and van Houte, 1975a; van Palenstein Helderman, 1975; Listgarten, 1976; van Palenstein Helderman and Rosman, 1976; Slots, 1976, 1977a,b; Fejerskov et al., 1977; Socransky, 1977). As the analysis of these communities has proceeded, the need for further studies on the taxonomy of oral bacteria has been recognized. In addition, it has become evident that details of the components and metabolic products of these localized communities must be better understood in order to explain the mechanisms of oral bacterial diseases (Bowen et al., 1976; Stiles et al., 1976; Lehner, 1977b; Minah and Loesche, 1977a,b; Nisengard, 1977).

In the following sections, the composition of the oral flora and its development and variation with habitat and in disease will be described as an introduction to further discussion of the physiology and biochemistry of the oral microbial community.

2.2. The Bacterial Community in the Oral Cavity of Humans

2.2.1. Genera

The mouth supports many bacterial genera (Table I). Some are unique to the mouth and might be regarded as autochthonous (Alexander, 1971; Savage, 1977) to this habitat. Among the best described of these are the gram-positive rods *Actinomyces*, *Bacterionema*, *Rothia*, and *Leptotrichia*, which represent autochthonous genera that have not been isolated from any other habitat

Table I. Bacterial Genera Found in the Mouth of Humans

Genera	Selected references[a]
Streptococcus	Colman and Williams (1972); Hardie and Bowden (1976b); Hardie and Marsh (1979)
Lactobacillus	Rogosa (1970); Edwardsson (1974); Sundquist and Carlsson (1974)
Neisseria	Berger and Wulf (1961); Berger (1967); Berger and Catlin (1975)
Haemophilus	Sims (1970); Killian (1976)
Arachnia	Edwardsson (1974); Slack and Gerencser (1975); Bowden and Hardie (1978)
Bifidobacterium	Edwardsson (1974); Scardovi and Crociani (1974); Slack and Gerencser (1975)
Actinomyces	Bowden and Hardie (1973, 1979); Holmberg and Nord (1975); Slack and Gerencser (1975); Fillery *et al.* (1979)
Eubacterium	Edwardsson (1974); Sundquist (1976); Hofstad and Skaug (1978)
Propionibacterium	Edwardsson (1974); Sundquist (1976); Bowden and Hardie (1978); Sanyal and Russell (1978)
Rothia	Slack and Gerencser (1975); Bowden and Hardie (1979)
Bacterionema	Slack and Gerencser (1975); Beighton and Miller (1977); Bowden and Hardie (1979)
Veillonella	Rogosa and Bishop (1964); Rogosa (1964); van Palenstein Helderman (1975)
Bacteroides	Holbrook and Duerden (1974); Williams *et al.* (1975); van Palenstein Helderman (1975); Shah *et al.* (1976); Sundquist (1976); Swindlehurst *et al.* (1978)
Fusobacterium	Hadi and Russell (1968); Sundquist (1976); Socransky (1977)
Campylobacter	van Palenstein Helderman (1975); Sundquist (1976)
Peptococcus	Sundquist (1976)
Peptostreptococcus	Sundquist (1976); Socransky *et al.* (1977)
Selenomonas	van Palenstein Helderman (1975); Sundquist (1976)
Eikenella	Socransky (1977)
Leptotrichia	Bowden and Hardie (1971)
Micrococcus	Bowden (1969)
Spirochaeta	Bowden and Hardie (1971); Loesche (1976b)

[a]The references are selected to provide some information on the genera in relation to the oral cavity.

(Bowden and Hardie, 1971, 1973, 1978; Holmberg and Nord, 1975; Slack and Gerencser, 1975). The other oral genera often have representative species elsewhere in the body. Specific details of the genera can also be found in *Bergey's Manual* (Buchanan and Gibbons, 1974).

Members of the Enterobacteriaceae, *Clostridium*, and *Corynebacterium sensu stricto* (Buchanan and Gibbons, 1974) have only been isolated from the mouth in exceptional circumstances, and it must be assumed that these genera are unable to compete effectively in the normal oral environment (Loesche *et al.*, 1972; Hardie and Bowden, 1974a). However, it should be noted that *Bacterionema* may be an oral representative of the genus *Corynebacterium* (Alshamony *et al.*, 1977; Bowden and Hardie, 1978).

Some of the genera can be regarded as more successful than others when measured by (a) their numbers in the community and (b) the range of oral habitats in which they can survive. Genera such as *Actinomyces* and *Streptococcus* can be isolated from most areas in the mouth and would therefore seem to occupy relatively wide niches (Hutchinson, 1965; Bowden *et al.*, 1978). Other genera are found in lower numbers and often grow in circumscribed oral environments, suggesting that they occupy narrower niches. Examples of these are the spirochetes (Listgarten and Lewis, 1967; Loesche, 1968) and some other gram-negative anaerobes (van Palenstein Helderman, 1975; Slots, 1976, 1977a,b; Socransky, 1977). However, as far as is known, no single genus is able to exclude all others from a habitat. In some cases, a genus will contribute significantly to a community and persist for long periods of time (Bowden *et al.*, 1975), but even in such defined habitats as carious dentine (Loesche and Syed, 1973; Edwardsson, 1974) or periodontal lesions (Socransky, 1977), several bacterial populations exist.

Many of the recent studies on oral bacteria have concentrated on *Streptococcus* and *Actinomyces*, species of which have featured as potential pathogens in caries and periodontal disease (Stiles *et al.*, 1976; Socransky, 1977). However, the gram-negative anaerobic genera are likely to be the subject of taxonomic studies in the future (Socransky, 1977). Despite the need to identify members of the oral bacterial communities, detailed taxonomic examinations of oral genera other than *Streptococcus* and *Actinomyces* are relatively few. Killian (1976) has published a definitive paper on *Haemophilus*, and the gram-positive rods other than *Actinomyces* have received some attention (Holmberg and Hallander, 1973; Holmberg and Nord, 1975; Fillery *et al.*, 1978). The studies of the genus *Streptococcus* have been concentrated on *S. mutans* (Stiles *et al.*, 1976), although a more general approach has been taken in some cases (Hardie and Bowden, 1976b; Hardie and Marsh, 1979). Genera such as *Eubacterium*, *Peptococcus*, and *Peptostreptococcus*, which are a taxonomic problem in other areas, have received little attention by oral microbiologists.

Table II. Bacteria Isolated from the Oral Cavity in Humans

Streptococcus	*Arachnia*	*Peptostreptococcus*
sanguis	propionica	anaerobius
salivarius	*Leptotrichia*	micros
mutans	buccalis	*Eikenella*
mitior	*Propionibacterium*	corrodens
milleri	acnes	*Fusobacterium*
mitis	freudenreichii	nucleatum
constellatus	jensenii	russi
intermedius	*Bifidobacterium*	*Selenomonas*
Actinomyces	eriksonii	sputigena
viscosus	dentium	*Campylobacter*
naeslundii	*Eubacterium*	sputorum
israelii	saburreum	*Treponema*
odontolyticus	alactolyticum	species
Lactobacillus	lentum	*Borrelia*
casei	*Neisseria*	species
acidophilus	flavescens	*Bacteroides*
salivarius	mucosa	melaninogenicus
plantarum	sicca	subsp. *asaccharolyticus*
fermentum	subflava	melaninogenicus
cellobiosus	*Veillonella*	subsp. *intermedius*
brevis	parvula	melaninogenicus
buchneri	alcalescens	subsp. *melaninogenicus*
catenoforme	*Peptococcus*	ochraceus
crispatus	species	ruminicola
Rothia	*Haemophilas*	oralis
dentocariosa	species	

2.2.2. Species

Most of the oral bacterial genera are composed of several species (Table II), although *Leptotrichia, Bacterionema, Rothia,* and *Arachnia* have only one accepted species each at this time (Bowden and Hardie, 1978). When definition of species within the oral genera is attempted, it becomes apparent that much information is lacking. Coupled with the problems of generic definition, the classification and identification of species populations in the mouth must be included as one of the major problems in the study of oral ecology. Even with the well-studied genera *Streptococcus* (Hardie and Marsh, 1979) and *Actinomyces* (Bowden and Hardie, 1973, 1978; Slack and Gerencser, 1975; Holmberg and Nord, 1975; Fillery *et al.,* 1978), problems still exist. The species *Streptococcus mitior* and *S. milleri* probably represent a range of bacteria, while *S. mutans* could be further divided by homology data (Coykendall, 1976). *Actinomyces viscosus* and *A. naeslundii* may be variants of the same species, although the animal strains of *A. viscosus* seem to merit separation (Fillery *et al.,* 1978).

Oral species of *Haemophilus* have been well described (Sims, 1970; Killian, 1976), and one species, *Haemophilus segnis*, seems to be uniquely associated with the mouth.

Gram-negative anaerobic rods are currently being classified into species in several laboratories because of their possible involvement in periodontal disease (Socransky, 1977). Recently, the well-established subspecies of *Bacteroides melaninogenicus* have been studied (Williams *et al.*, 1975; Shah *et al.*, 1976; Swindlehurst *et al.*, 1977; Williams and Shah, 1979), with the result that *B. melaninogenicus* subsp. *asaccharolyticus* has been given a new species designation, *B. asaccharolyticus* (Finegold and Barnes, 1977). In addition *B. melaninogenicus* subsp. *intermedius* of oral origin, although phenotypically identical to *B. melaninogenicus* subsp. *intermedius* of fecal origin, shows no DNA homology with the fecal strains (H. Shah, personal communication from J. Johnson). This may indicate a very fine and distinct ecological relationship between these organisms and their habitats which is not detected by phenotypic tests. The gram-negative nonpigmented anaerobic rods in the mouth are not as well understood and have received less attention. It seems likely that fecal species, e.g., *Bacteroides fragilis*, are not common in the mouth of man, although strains resembling *Bacteroides ruminicola* have been isolated (McKee and Shah, 1979). However, the oral strains of *B. ruminicola* do show some variation from the rumen isolates in that they are less sensitive to O_2 and, unlike strains from the rumen, they can be easily cultivated in anaerobic jars using bench techniques (H. Shah, personal communication). A further problem with the nonpigmented strains is the validity of the species *Bacteroides oralis* (Finegold and Barnes, 1977) and the relationship of *"Capnocytophaga"* isolates to *Bacteroides ochraceus* (Socransky, 1977; Sasaki *et al.*, 1978; Maeder *et al.*, 1978). Undoubtedly, more information will accumulate on the species of other gram-negative, anaerobic rod-shaped bacteria Socransky, 1977; van Palenstein Helderman, 1975; Slots, 1976, 1977a,b).

Among the gram-negative cocci, the work of Rogosa (1965) remains the only extensive study of the taxonomy of the genus *Veillonella*. Two species, *Veillonella alcalescens* and *V. parvula*, are present in the mouth, although there has been a recent proposal, based on DNA homology data, that these two species should be combined (Holdeman and Moore, 1977). *Veillonella* species utilize lactate and may be of considerable importance in the food chain in dental plaque, and as such they deserve further study.

Although *Neisseria* are common oral organisms, they are difficult to place in species. This is evidenced by the description of the commensal species of *Neisseria* in the most recent edition of *Bergey's Manual* (Buchanan and Gibbons, 1974). Ten species of uncertain status are listed, and of those accepted, only *Neisseria flavescens* is shown with a type strain. The oral *Neisseria* have been examined by Berger and Wulf (1961), Fahr and Berger (1975), and Berger and

Catlin (1975), and although some of the more sensitive taxonomic techniques have been applied to the *Neisseria* (Bovre, 1967; Brooks *et al.*, 1971; Johnson and Sneath, 1973; Fox and McClain, 1974; Holten, 1974; Jantzen *et al.*, 1975; Russell *et al.*, 1975; Russell and MacDonald, 1976), confident identification of oral strains is not possible.

Oral gram-positive rods have been examined in some detail, and as several of the genera are accepted as monotypic, identification of the species does not present a problem. *Rothia* may have more than one species (Slack and Gerencser, 1975; Bowden and Hardie, 1978), and oral *Bifidobacterium* have not been subjected to much study, although species have been described (Edwardsson, 1974; Scardovi and Crociani, 1974; Holmberg, 1976; Bowden and Hardie, 1978). *Lactobacillus* species are among the best-described species of oral gram-positive rods (Rogosa, 1970; Edwardsson, 1974), and procedures are available for their identification. Studies have been made on oral strains of *Propionibacterium* (Sanyal and Russell, 1978), and with the data available (Cummins and Johnson, 1974; Edwardsson, 1974; Cummins, 1975; Holmberg, 1976), careful identification should be possible. Probably the most difficult of the oral rods to classify into species will be members of *Eubacterium*, although *Eubacterium saburreum* and *E. alactolyticum* have been described (Edwardsson, 1974; Sundquist, 1976; Holdeman and Moore, 1977; Hofstad and Skaug, 1978), and other species almost certainly exist (Sundquist, 1976). Together with *Eubacterium*, the anaerobic gram-positive cocci and the oral spirochetes represent the least understood of the bacterial species in the mouth. The Virginia Polytechnic Institute *Anaerobe Laboratory Manual* (Holdeman and Moore, 1977) is valuable for the identification of many of the oral anaerobes, but it is necessary for fresh isolates to be carefully examined and new taxonomic methods employed in order to aid the identification of oral bacteria. In addition, isolation techniques may have to be developed, as it is unlikely that all species of oral bacteria can be cultivated at this time.

Further studies on the taxonomy of oral bacterial genera could reveal valuable data on the ecology of their various species. Detailed information on the characteristics of the species of *Streptococcus* and *Actinomyces* has resulted in a better understanding of some of the ecological parameters in the mouth. Knowledge of the characteristics of species within other genera could extend our appreciation of the ecology of the oral cavity.

2.3. The Bacterial Community in the Oral Cavity of Animals

Few detailed studies have been made on the oral flora of animals, and several of these concern the primates (Cornick and Bowen, 1971; Brown *et al.*, 1973; Dent *et al.*, 1976, 1978). Some information is available on rodents (Jordan and

Keyes, 1964; Coykendall *et al.*, 1974; Dvarskas and Coykendall, 1975; Huxley, 1976), although in most studies the samples were taken from animals in captivity and often undergoing experimentation (Mikx *et al.*, 1972). Unfortunately, almost no examinations have been made on animals in the wild, since it might be expected that these would better reflect the effects of varying oral conditions on the composition of the bacterial community.

Some authors have considered animal species other than primates. Plaque from the tooth surfaces of a total of 33 animals in the London Zoo representing 22 animal species was cultured and a wide range of bacteria isolated and characterized (Table III) (Dent *et al.*, 1976, 1978). The results of the identification of the streptococcal isolates suggest that animals consuming a diet relatively high in carbohydrate have *S. mutans* and *Streptococcus sanguis* as components of their dental plaque (Dent *et al.*, 1978). A further observation was that certain genera, including *Actinomyces* and *Streptococcus*, were isolated from almost all of the animals examined. This suggests that these genera may be components of a "basic" plaque community common to many animals and man (Bowden *et al.*, 1978). Selection of specific bacterial species from this "basic" community would occur through changes in the oral environment; e.g., the consumption of a carbohydrate diet by the animals would result in the presence of glucan-producing streptococci (see Section 3). Selection of *A. viscosus* in rats by glucose

Table III. Bacteria Present in the Plaque of Animals[a,b]

Organism	Herbivores (yak, lamb, axis deer, bennet wallaby)	Carnivores (singing dingo, black bear)	Apes and monkeys (patas monkey, spider monkey, baboon, gorilla)	Other animals (Indian fruit bat, spiny tenrec, bushbaby, red mantled tamarin)
Streptococcus	+	+	+	±
Actinomyces	+	+	+	±
Neisseria	±	∓	±	∓
Corynebacterium	±	0	∓	0
Bacteroides	0	0	±	0
A. viscosus	∓	0	∓	0
B. matruchotii	∓	0	±	∓
S. bovis	±	0	0	∓
S. mutans	0	0	∓	∓
S. sanguis	0	0	+	+
S. salivarius	0	0	±	0

[a] Data taken from Dent *et al.* (1978), Bowden *et al.* (1978), and V. Dent (personal communication).
[b] +, Present in all animals examined; ±, usually isolated; ∓, only occasionally isolated; 0, not isolated from animals.

feeding has been described by van der Hoeven (1974). Recently, Beighton and Miller (1977) cultivated plaque from macropods at the Melbourne Zoo. They found that gram-positive rods constituted a major component of the plaque flora. *A. viscosus* was isolated from all 12 of the animals, while *Bacterionema matruchotii* was isolated from 11. Enterococci were the most common streptococci isolated and generally represented 12% of the cultivable flora. As higher counts were obtained under aerobic rather than anaerobic conditions and only low levels of anaerobes were isolated, it was assumed that higher O_2 tensions existed in macropod plaque as compared to human plaque.

Primates seem to have a flora closer to that of man than do other animals (Brown *et al.*, 1973; Dent *et al.*, 1976, 1978), and for this reason they should be the most useful experimental animals for models of caries and periodontal disease. Apart from this, the study of the flora of animals could reveal some of the factors controlling the composition of the bacterial communities in the mouth. The results of the work of Beighton and Miller (1977) seem to show that under more aerobic conditions the composition of the communities on the teeth is relatively simple. Some variation in the streptococcal species present in plaque from different animals was also noted by Dent *et al.* (1978); e.g., *Streptococcus bovis* was restricted to herbivores.

2.4. Variations in the Bacterial Community with Habitat

2.4.1. Tooth Surfaces

Those bacteria which colonize the surface of the teeth make up one of the best-studied oral communities. These aggregations of bacteria are generally described as dental plaque. Previous to the 1960s, little interest had focused on dental plaque. It was relatively ill defined and assumed by many to be an accumulation of food debris. However, early studies demonstrated that plaque was composed predominantly of bacteria, their extracellular products, and salivary components (Slack and Bowden, 1965; Howell *et al.*, 1965; Ritz, 1967; Newman and Poole, 1974; Guggenheim, 1976). By 1970, sufficient data on dental plaque had accumulated to warrant a symposium on its composition and biochemistry (McHugh, 1970). Since that time, extensive studies have been made on plaque structure, composition, development, interactions, products, and pathogenicity (Stiles *et al.*, 1976).

The tooth surface presents several environments for the growth of bacteria. In general, there are the smooth surfaces of the tooth below (subgingival) and above (supragingival) the gum margin. The biting (occlusal) surfaces of the teeth provide unique retentive areas in the fissures of the enamel. Some subdivision of the smooth supragingival surfaces of the enamel is also made. Areas between two adjacent teeth are described as approximal, while the surface against the cheek is buccal and that against the tongue lingual. Among the various surfaces, those of the fissures and approximal areas can be regarded as the most retentive, both providing physical protection for microbial communities.

Development of a bacterial community on the enamel surface depends upon the ability of individuals to localize, adhere, and grow in the presence of several factors which must be regarded as antagonistic to colonization (Gibbons and van Houte, 1975a; van Houte, 1976). Detailed studies on the mechanisms of colonization of oral surfaces have an obvious relationship to the control of oral disease, and extensive examinations have been made on this topic. (For a discussion of the mechanisms involved in the initial colonization of the tooth surface, see Section 3. Similarly, bacterial interactions play an important role in the development of microbial communities on the tooth enamel and will be discussed in Section 5.)

2.4.1a. Supragingival Dental Plaque

Smooth Surfaces. The development of the bacterial community on an enamel surface follows a general pattern, which has been demonstrated by microbiological and histological studies on natural enamel and artificial devices (Howell *et al.*, 1965; Slack and Bowden, 1965; Ritz, 1967; Saxton, 1973, 1975a,b; Listgarten *et al.*, 1975; Syed *et al.*, 1975; Hardie and Bowden, 1976a; Guggenheim, 1976; Socransky *et al.*, 1977).

Dental plaque shows the characteristics expected of a developing microbial community (Alexander, 1971), with a gradual increase and change in the bacterial species present and with some organisms persisting for considerable periods of time (Bowden *et al.*, 1975). After some time, the communities can reach a "climax" state of relative stability (Alexander, 1971), although on a tooth surface, it might be expected that this would occur most easily in the relatively protected areas, e.g., approximal sites. The other surfaces of the teeth may be more open to dramatic changes in the environment such as those resulting from oral-hygiene procedures. As would be expected in a community, organisms occupy specific niches. For example, the initial studies of Mikx and van der Hoeven and their associates have indicated that *Veillonella* may be dependent on lactate for survival in plaque, although this may not always be the case (Mikx *et al.*, 1976a,b; van der Hoeven *et al.*, 1978). Also, as mentioned previously, it has been shown that *S. sanguis* could satisfy a nutritional requirement of *S. mutans* (Carlsson, 1971). However, there is little information on the precise parameters of a niche for any oral bacteria.

Cleaned tooth enamel rapidly acquires a layer of deposited salivary components or "pellicle" (Tinanoff *et al.*, 1976; Rolla, 1976), so that those bacteria which colonize the surface interact with this deposit as well as with free enamel. The early studies on plaque development were in general agreement that strains

of *Streptococcus* and *Neisseria* species colonized the tooth surface within 24 hr
(Slack and Bowden, 1965; Ritz, 1967) and that the community increased in
complexity over a period of up to 14 days (Hardie and Bowden, 1976a; Guggenheim, 1976). The increase in complexity is associated with the appearance of
anaerobes and a lowering of the E_h of plaque (Kenney and Ash, 1969), and such
a progression could be regarded as an autogenic succession (Alexander, 1971).
The details of the microorganisms detected at the various stages of plaque development are given in Table IV.

The most recent study of the development of plaque on an enamel surface
bears out the possibility that other bacteria can be involved in early colonization
(Socransky *et al.*, 1977). Socransky and his co-workers examined the organisms
on enamel as early as 5 min after prophylaxis. Although *S. sanguis* was the predominant organism associated with early colonization, *A. viscosus*, *A. naeslundii*,
and *Peptostreptococcus* were present after 1 hr. These results were taken from
a single site on a tooth, but it was selected with care to represent an area of
enamel which would be colonized by only the most retentive and hardy oral
strains. Different patterns of colonization might be expected in more protected
areas. One interesting aspect of the study was the isolation of organisms which
produced heterogeneous colonial morphology and distinctly different cellular
morphology. The strains were designated "suspected pairs," and as the variations
existed during ten sequential transfers of single colonies, it was not clear whether
they were mixed cultures or two mutually dependent bacteria. This result shows
that several bacteria are involved in initial colonization and supports the suggestion that a range of mechanisms could be expected to be involved in adherence
and retention (van Houte, 1976). These observations should encourage the

Table IV. Bacteria Detected at Various Stages of Plaque Development on
Cleaned Solid Surfaces in the Mouth[a]

Time after cleaning	Species present
0–15 min	*S. sanguis*, *S. salivarius*, *A. viscosus*, "corynebacterium"
1–18 hr	*S. sanguis*, *S. mitis*, *S. epidermidis*, *A. viscosus*, *Peptococcus* sp.
24–48 hr	*S. sanguis*, *S. salivarius*, *A. viscosus*, *R. dentocariosa*, *L. casei*, *Veillonella* sp., *Fusobacterium* sp., *Neisseria* sp.
3–5 days	*S. sanguis*, *S. salivarius*, *A. viscosus*, *A. naeslundii*, *A. odontolyticus*, *R. dentocariosa*, *L. buccalis*, *E. saburreum*, *A. israelii*, *B. melaninogenicus*, *Neisseria* sp., *Veillonella* sp., *Lactobacillus* sp.
6–14 days	At this time, the plaque reaches its most complex community and includes the range of bacteria described in Section 2.4.1a and Table VI.

[a]Data taken from Slack and Bowden (1965), Ritz (1967), Hardie and Bowden (1976a),
Socransky *et al.* (1977), and Syed and Loesche (1978).

examination of the basis of such mechanisms using well-defined physicochemical systems (Rutter and Abbott, 1978).

Observations on the composition of established communities followed those on development. Different communities were found to be present on different areas of the same tooth (Hardie and Bowden, 1974a), and this finding was of considerable significance as caries and periodontal disease are often localized. Continuing on this theme, Bowden and co-workers (1975) removed well-defined samples from approximal areas and proposed some parameters for the characterization of the bacterial communities: (a) the qualitative bacterial composition in terms of genera and species, (b) the quantitative bacterial composition, (c) the stability of the community over a period of time, and (d) the persistence of specific bacterial species.

When these parameters were applied to samples from approximal areas in a longitudinal study, it was seen that some habitats possessed a persistent and unique level of a specific population within the community (Bowden et al., 1976; Hardie et al., 1977). Stable and unstable communities were also demonstrated, and certain bacteria, such as A. viscosus and A. naeslundii, were shown to colonize the approximal area better than others. Furthermore, it was noted that populations of S. mutans, usually present in low numbers, could increase in the community and achieve prominence for periods of up to 18 months. The very localized nature of these populations was demonstrated by the fact that they did not occur in an identical area on the opposite side of the mouth (Bowden et al., 1976). Increases in the numbers of S. mutans and the localized nature of their populations have also been noted by other workers (Gibbons et al., 1974). Examination of more recent data has shown that S. mutans may be one of the few oral organisms which is able to increase and achieve stable prominence in this way in humans, whereas S. mitior, S. sanguis, Veillonella sp., Lactobacillus sp., and B. melaninogenicus seldom increased for prolonged periods (Marsh et al., 1978) (Table V).

The bacteria present in "established" approximal plaque in humans can be grouped together depending on their contribution to the community (Table VI). It should be noted, however, that the organisms may be distributed differently on exposed smooth surfaces where, for example, S. sanguis would be expected to be a prominent organism (Gibbons and van Houte, 1975a; Socransky, 1977).

As has been mentioned previously, comparison of the organisms mentioned above to those found in animals indicates that some are able to colonize the tooth surfaces independent of diet in differing environmental conditions. In addition, bacteria can be isolated from the supragingival plaque of primitive man (Hardie and Bowden, 1974a) which are common to those of people consuming a Western diet. Recent examination of plaque from tribesmen in New Guinea has revealed a complex flora similar to that of people in the Western World (Schamschula et al., 1978). The emergence of some species of bacteria in the

Table V. Stable Increases in Populations on the Tooth Surface
Leading to Prominence in the Approximal Community

Contribution to total viable count (%)	Number of sites ($n = 64^a$) showing stable[b] increases in the bacterial population				
	S. mutans	*S. sanguis/mitior*	*B. mel.*[c]	*Veill.*[c]	*Lact.*[c]
<5	3	1	1	0	2
5–15	5	0	1	0	1
15–25	4	0	0	0	0
>25	2	0	1	1	0
Totals	14	1	3	1	3

[a]Data taken from Marsh *et al.* (1978).
[b]The increases persisted for at least 12 months at the levels shown.
[c]Abbreviations: *B. mel., B. melaninogenicus; Veill., Veillonella* sp.; *Lact., Lactobacillus* sp.

plaque community is probably dictated in part by diet; for example, *S. mutans* establishment is favored by sucrose (van Houte, 1976) and *Veillonella* may be associated with lactate (Mikx *et al.*, 1976b). Some evidence for selection of this type in animals is provided by Dent *et al.* (1978).

The bacterial inhabitants of supragingival plaque must reflect the environment in the specific habitat. It is not known how rapidly changes in the environment would be reflected by modifications in the flora. If bacteria autochthonous to supragingival plaque occupy wide niches and make up a high proportion of the community, extensive changes may be produced only with difficulty; e.g., *Actinomyces* species might survive extremes of environment. Organisms with

Table VI. Examples of the Contributions of Various Bacteria to the
Approximal Plaque Community

Regularly isolated		Isolated in localized situations
High numbers	Low numbers	
Bacteroides sp.	*Neisseria* sp.	*A. israelii*
Streptococcus sp.	*Fusobacterium* sp.	*A. odontolyticus*
A. viscosus	*S. sanguis*	*B. matruchotii*
A. naeslundii	*S. mitior*	*L. buccalis*
V. alcalescens	*S. milleri*	*B. melaninogenicus* subsp. *asaccharolyticus*
Haemophilus sp.	*R. dentocariosa*	*B. melaninogenicus* subsp. *intermedius*
		E. alactolyticum
		E. saburreum
		S. salivarius
		Lactobacillus sp.
		A. propionica

narrower niches might be expected to reflect changes in the environment more readily, and the increases in *S. mutans* populations (Table V) may represent such a response to an environmental change. However, in general terms, areas of supragingival plaque are composed of the same organisms, although they may be present in different proportions. This is to be expected, as members of the community have been selected by their ability to survive in this habitat.

Occlusal Surfaces. The occlusal surfaces of the teeth present a different structural habitat for bacterial colonization in that the fissures provide protected retentive areas for growth. Study of the microbiology of the bacterial community in fissures is difficult due to the lack of access, and, consequently, several of the examinations have been made on artificial fissures (Löe *et al.*, 1973; Theilade *et al.*, 1974; Minah and Bowman, 1978; Svanberg and Loesche, 1978a,b). Studies by Theilade *et al.* (1974) and Thott *et al.* (1974) recognized that the development of a bacterial community in an artificial fissure could vary from that in a natural one. Consequently, these workers used fissures from non-erupted teeth held in position *"in vivo."*

The results of these studies of the fissure communities showed that there was little difference between those in artificial or normal fissures. Species present in early development include *S. mutans*, *S. sanguis*, and *S. salivarius*, together with strains of *Lactobacillus*, *Haemophilus*, and *Veillonella* (Theilade *et al.*, 1973; Theilade *et al.*, 1974; Thott *et al.*, 1974). Exposure of the fissures in the mouth for up to 260 days did not cause any greater complexity (Fejerskov *et al.*, 1977). However, the numbers of *S. mutans* in the communities did increase in older fissure plaque, and it seems possible that this retentive site is a major habitat for this organism (Gibbons and van Houte, 1975a; Fejerskov *et al.*, 1977; Loesche *et al.*, 1978). Little information is available on the types and numbers of *Actinomyces* in fissure plaque, but it could be that they are not as significant in fissures as they are in the approximal area.

Very few studies have been made of the bacterial colonization of the fissures of animal teeth (Huxley, 1972, 1976; Mikx *et al.*, 1975a,b).

Histological studies of fissure plaque have been ably reviewed by Guggenheim (1976). In general, the observations made under the electron microscope support the microbiological results, although electron microscopy gives the impression that the majority of the cells in mature fissures are nonviable. This has not been borne out by the cultural techniques.

Subgingival Surfaces. A detailed description of the relationship between the tooth and the gingiva is given by Schroeder (1977). In most people, the area between the tooth and the gum margin (gingiva) forms a small pocket or sulcus. Schroeder points out that while this sulcus is not present in gnotobiotic animals or in subjects undergoing intensive tooth cleaning, it is commonly found in conventional animals and normal people. Thus, a pocket of varying depth will exist between the tooth and gingiva, and a decision as to its "diseased" or healthy state is generally determined by one or more clinical indices (Russell, 1956; Löe and Silness, 1963). The state of the gingiva and the degree of inflammation have some bearing on the microbial community, as these factors influence the rate of transport of leucocytes and tissue fluids into the sulcus (Schroeder, 1977). The fluid entering the sulcus contains serum immunoglobulins (Brandtzaeg and Tolo, 1977), complement (Wilton, 1977), and leucocytes Renggli, 1977), which probably enter as the result of chemotaxis (Schroeder, 1977). The bacterial community found in association with the enamel of the "pocket" develops against these antagonistic factors.

Some studies on the bacteriology of pockets that were regarded as normal have been made (Beveridge and Goldner, 1973; Williams *et al.*, 1976; Newman and Socransky, 1977; Slots, 1977a). Slots (1977a) regarded a sulcus of less than 3-mm depth with healthy gingiva as normal. Samples from seven individuals were cultivated on selective and nonselective media using a roll-tube technique. The results showed that the communities were complex. Gram-positive anaerobic and facultative rods composed a mean of 44.6% of the cultivable organisms and were the major population in three of the samples. The rods were predominantly *Actinomyces*, including *A. viscosus*, *A. naeslundii*, and *A. israelii*; *Arachnia propionica* and *Rothia dentocariosa* were isolated in low numbers. *S. sanguis* and *Streptococcus mitis* were identified from the samples, and this genus made up a mean of 39.6% of the cultivable organisms. *Streptococcus* were the predominant organisms in four of the samples. *Bacteroides* and *Fusobacterium* constituted a mean of 12.7% of the communities, and *Veillonella* and *Peptococcus* were found only in low numbers. Newman and Socransky (1977) used a relatively sophisticated technique to examine the flora of normal sites in their subjects. The results agreed in the essential details with those of Slots (1977a). The predominant flora was gram positive and included members of *Streptococcus* and *Actinomyces*.

A comparison of the figures given above to those for the communities of supragingival plaque shows that the populations, with the possible exception of *Veillonella*, are similar. This similarity is borne out by the isolation of the same species from supragingival plaque and the flora of the "normal" sulcus. Also in common with supragingival plaque, there is considerable variation in the numbers of the various species present in different samples. Thus, it seems very likely that the flora detected in the "normal" gingival sulcus is simply an extension of the supragingival plaque in that area.

2.4.2. Tissue Surfaces

Relatively few studies have been concerned with the detailed flora of oral habitats other than those intimately associated with the tooth surface (Hardie and Bowden, 1974a). Those studies which have examined other mucosal surfaces have been concerned predominantly with the specific adherence of certain *Streptococcus* species (Gibbons and van Houte, 1975a).

2.4.2a. *The Tongue.* The tongue possesses a specific flora, with *S. salivarius* and *Veillonella* being prominent members of the community (Gibbons and van Houte, 1975a). In contrast to the communities on the teeth, *S. sanguis* and *Actinomyces* species are seldom isolated. The tongue therefore provides a good demonstration of the individuality of the communities associated with the major habitats in the mouth. This individuality means that despite the close physical association of the tongue and the teeth, each habitat carries its own autochthonous species. These may play an allochthonous role in other ecosystems. *S. salivarius* is one such organism, being common on the tongue but only occasionally being isolated from plaque. Another organism commonly associated with the tongue, but rarely found in plaque, is *Micrococcus mucilaginosus* (Kocur *et al.*, 1971). This organism produces an extensive extracellular polysaccharide (EPS) (Bowden, 1969) independent of the presence of carbohydrate. It is possible that this slime could play a role in lubrication of the tongue.

There have been no detailed studies of the flora of the tongue using modern isolation and identification techniques. Undoubtedly, the flora is complex, but due to the interest in the bacteriology of caries and periodontal disease, studies of the flora of the tongue and areas in the mouth not associated with the teeth have not been attempted.

2.5. Variations in the Bacterial Community with Environmental Changes

It is to be expected that changes in the local environment would be reflected in the composition of the bacterial communities. However, although it is relatively easy to see that a reduction in saliva or an extreme dietary change causes a variation in the environment, other situations are less clear. For example, in oral disease, changes in the environment do occur, but it is not obvious at which point these begin to influence the bacterial communities. In diseases such as caries or periodontal disease, it is difficult to separate the change in the community which produces disease from that which results from tissue destruction. For these reasons, the best examples of the effect of environment on oral bacterial communities are seen with variation in saliva flow and high consumption of refined carbohydrate. Although the bacterial populations in oral disease have been described, the precise etiology is not known.

2.5.1. Saliva

Saliva plays a major role in the control of the accumulation of microorganisms on oral surfaces (Geddes and Jenkins, 1974; van Houte, 1976). It does this by acting as a carrier for a wide variety of antibacterial agents (see review by

Mandel, 1978) including immunoglobulin, lysozyme, lactoferrin (Cole *et al.*, 1976), lactoperoxidase (Hoogendoorn, 1976), and high-molecular-weight glycoproteins. The glycoproteins have the capacity to aggregate bacteria and may also have an affinity for hydroxyapatite (Gibbons and van Houte, 1975a; Ericson *et al.*, 1976; van Houte, 1976). Extensive studies have been carried out on the glycoproteins which can show some degree of specificity in their reactions (Ericson *et al.*, 1976; Gibbons and Qureshi, 1976; Levine *et al.*, 1978).

The bacteria present in saliva are generally assumed to be derived from the oral surfaces, particularly the tongue, as a result of its cleaning activity (Hardie and Bowden, 1974a; Gibbons and van Houte, 1975a). Perhaps the best indication of the influence of saliva on the oral flora can be gained by examining data from patients in whom the salivary flow is absent or reduced. In two papers, Dreizen and Brown (1976) and Brown *et al.* (1976) describe the changes in disease and the oral flora in cases of xerostomia induced by irradiation of the salivary glands during radiotherapy. Loss of saliva flow caused plaque accumulation and a rapid increase in caries activity.

The analysis of the microbial communities associated with these patients (Brown *et al.*, 1976) showed that, in general, an acidogenic flora was produced, with a change in the ratios of certain populations in the communities. Increases were detected in the "anaerobic diphtheroids," *S. mutans*, and *Lactobacillus* species. Thus, in xerostomia, a reduction in saliva flow encourages an increase in numbers of acidogenic bacteria, and with normal diets, a low-pH environment develops (Edgar, 1976) this environment favors aciduric bacteria and promotes an active caries situation. Control of bacterial plaque by oral-hygiene procedures, together with diet control and the application of fluoride (see Section 4.6), removes this advantage, and the population in the community returns to normal.

2.5.2. Diet

The effect of a specific diet on the composition of the oral bacterial communities is little understood. Measurements have been made, and these generally concern the consumption of sucrose or other refined carbohydrates (Bibby, 1976). Examinations of the flora from people of different races consuming their natural diets suggest that the oral bacterial communities are similar (Schamschula *et al.*, 1978). It has also been shown on several occasions that consumption of food is not essential for plaque formation. The bacterial communities develop even when animals or people are fed by tube (Bibby, 1976). Bibby (1976) reviewed several of these studies of the effect of diet on plaque and concluded that "the bacterial population of the mouth and of plaques is the product of evolutionary process. It is apparently dependent to a much greater extent upon the natural environment provided by the mouth than by the transient presence of foodstuffs therein."

This does not conflict with the results of the ingestion of high levels of sucrose causing an increase in the numbers of some populations. High carbohydrate consumption producing increases in *S. mutans* and *Lactobacillus* species is, perhaps, the only well-documented example of the effect of diet on oral communities. Almost no information is available, or sought, on the effect of high-protein or other diets. The significance attached to the effect of sucrose intake on the oral flora depends to some extent on the viewpoint of those examining the data. To the dentist, it is a very significant factor in the caries process; however, as an ecological phenomenon, it is just one example of changes introduced into oral communities by diet.

It seems reasonable to propose, therefore, that the oral bacterial communities in their various habitats seem able to compensate for natural variations in the diet. However, in the case of a high intake of carbohydrate, the acid conditions produced in localized habitats favor the increase of certain populations. In addition, sucrose ingestion may provide an advantage to those bacteria capable of producing extracellular glucans (see Section 3) (Gibbons and van Houte, 1975a; van Houte, 1976).

2.5.3. Immunological Factors

The mouth is provided with both humoral and cellular immune systems which could influence the composition of the microbial communities (Brandtzaeg, 1976; Brandtzaeg and Tolo, 1977; Lehner, 1977a; Renggli, 1977; McGhee *et al.*, 1978). The humoral mechanisms include local immunoglobulin production in the gingiva (Brandtzaeg and Tolo, 1977), secretion in saliva (Brandtzaeg, 1976), and the passage of serum immunoglobulins and complement components via the gingival sulcus (Wilton, 1977). Elements of cellular immunity are also present in the gingival sulcus (Lehner, 1977b; Schroeder, 1977; Renggli, 1977) and include polymorphonuclear leucocytes, lymphocytes, and monocytes. Thus, the bacterial communities growing in close association with the tooth, particularly in the gingival area, are exposed to the possibility of some immunological control. Despite this potential control, it is evident that many bacteria are capable of colonizing the surfaces in the mouth and can be found as constant components of the communities. This suggests that the autochthonous oral bacteria are either unaffected by, not subjected to, or able to avoid immune mechanisms. Effective suppression of some bacteria could be indicated by their absence from the mouth. Shedlofsky and Freter (1974) have proposed that control of the populations among the intestinal flora results from a combination of immune mechanisms and bacterial antagonisms, and it seems reasonable to propose a similar situation for the mouth.

Our knowledge of the immune responses of the host to oral bacteria derives from three major sources: (a) studies with pure antigen and bacteria, (b) attempts to control a population by immunological means as a basis for vaccination, and (c) measurements of responses to bacterial antigens in periodontal disease.

There is little doubt that antibody production can be stimulated by antigens associated with the gingival margin or oral mucosa. Early studies with rabbits (Rizzo and Mitchell, 1966; Berglund *et al.*, 1969), guinea pigs (Wilton, 1969), and monkeys (Ranney, 1970) showed that antigens at the gingival margin promoted antibody responses. Berglund *et al.* (1969) demonstrated that lipopolysaccharide from *Escherichia coli* given in low doses stimulated local lymph nodes, and higher doses produced a systemic response. In humans, several studies have detected circulating antibody to various oral bacterial antigens (Evans *et al.*, 1966; Kristoffersen and Hofstad, 1970; Wilton *et al.*, 1971; Hofstad, 1974; Gilmour and Nisengard, 1974; Challacombe and Lehner, 1976; Williams *et al.*, 1976; Orstavik and Brandtzaeg, 1977). Serum immunoglobulins reach the mouth via the gingival sulcus, and it has been shown recently that IgG antibodies with specificity for *V. alcalescens, A. viscosus,* and *S. mutans* are present in washings of the gingival area (Wilton, 1977).

Together with the serum immunoglobulins, the saliva contains secretory IgA, which provides the most commonly accepted protective antibody of mucosal surfaces (Herremans, 1974; Brandtzaeg, 1976; Mestecky, 1976; Brandtzaeg and Tolo, 1977). As with the serum immunoglobulins, in humans, salivary antibodies can be detected with activity against oral bacteria (Brandtzaeg *et al.*, 1968; Sirisinha, 1970; Sirisinha and Charupatana, 1971; Wilton *et al.*, 1971; Challacombe and Lehner, 1976). Sirisinha and Charupatana (1971) have presented evidence that salivary antibodies are directed in some cases against different indigenous bacteria from those in serum. These authors showed that parotid IgA, and the IgA in whole saliva, carried activity to *S. mitis, S. salivarius,* and enterococci and that parotid, but not serum, IgA, also showed activity against *Veillonella*. Antibody *E. coli* was found only in serum IgG, IgM, and IgA, not in parotid IgA. This suggests some local production of antibody to specific bacteria.

Although it is evident that antibody reacting with oral bacteria can be detected in serum and saliva, it cannot be said with any degree of certainty that these oral bacteria triggered production. Cross-reacting antibody could be produced as a result of stimulation by related organisms elsewhere in the intestinal tract. Of the antibodies described, the IgA in parotid saliva reacting with *Veillonella* is most likely to be the result of local stimulation via the mouth. The IgG antibody to *A. viscosus* (Wilton, 1977) in gingival fluid may also derive from oral stimulation, as *Actinomyces* are unique to the mouth. Another focus for the contact between cells of the immune system and oral organisms could be the tonsils (Brandtzaeg, 1976), where several species of oral bacteria can be detected (Blank and Georg, 1968).

Antibody could control populations of oral bacteria either by the inhibition of colonization by reactivity with surface receptors (Gibbons and van Houte, 1975a; Evans *et al.*, 1975) or as opsonins encouraging phagocytosis (Renggli, 1977; Wilton, 1977). One of the major questions in consideration of humoral control of the oral flora is, why do autochthonous bacteria survive in the mouth? Gibbons and van Houte (1975a) suggest that oral streptococci survive by antigenic variation, their antigenic makeup changing to avoid reactivity with antibody (Bratthall and Gibbons, 1975). This is in keeping with some observations on pathogens (Gibbons and van Houte, 1975a; Miller *et al.*, 1972), but the phenomenon of antigenic variation has not been explored in detail for members of the autochthonous flora. Indeed, it is possible that animals respond less readily to antigens from their indigenous flora than to organisms from the external environment (Foo and Lee, 1974; Berg and Savage, 1975). The importance of the indigenous flora in the stimulation of the natural antibody content of serum has been studied by several authors (Springer and Horton, 1969; Brown and Lee, 1974; Foo *et al.*, 1974), and these antibodies are often of the IgM class. It is possible that bacteria that are truly autochthonous are not readily recognized by the immune system. In contrast, organisms that are attempting to establish within the community might be recognized more easily. In the latter case, antigenic variation could be one device for successful establishment. In studies using gnotobiotes, these animals may place a greater immune pressure on the bacteria with which they are infected than normal animals. However, even gnotobiotic mice respond differently to monoassociation with bacteria indigenous to the mouse than with nonindigenous strains. One reason for this could be that the indigenous bacteria share antigens with the host (Foo and Lee, 1974).

Thus, the flora that is considered normal for a habitat could survive the pressures of normal immune systems either by having antigens cross-reacting with the host or by so changing their surface characteristics that they are able to avoid recognition for periods of time. The fact that the indigenous flora survives the immune system is of considerable importance when immunological control of an oral population is proposed. It may be that the effectiveness of such measures depends to some extent on the true relationship of the bacteria to its host. Organisms which could be regarded as "invaders" might be more susceptible than those which are autochthonous.

When vaccination against caries is proposed, the bacterial species selected for control is *S. mutans* (Bowen *et al.*, 1976; Genco, 1976). This organism must be considered a potential pathogen in caries (Gibbons and van Houte, 1975a), although it may not be the sole etiologic agent (Bowden *et al.*, 1976; Hardie *et al.*, 1977). However, the results of the attempts at control of *S. mutans* in animals are the only data on the effects of humoral mechanisms on a member of the oral flora. Generally, the studies can be divided into two groups: (a) those that concern the role of secretory IgA antibody induced either by local stimulation or via the gut and (b) those that concern the stimulation of serum antibodies which enter the mouth via the gingival sulcus. Both of these systems have met with some success in reducing the extent of caries in experimental animals.

The majority of the experiments on the stimulation of secretory IgA have been carried out on rodents, both normal and gnotobiotic. Oral immunization using killed whole cells has been shown to be effective (McGhee *et al.*, 1975; Michalek *et al.*, 1976a, 1978) and to produce secretory immunoglobulins in the absence of serum antibody. In common with other studies, gnotobiotic animals responded to a lower dose of antigen than those carrying a commensal flora (Michalek *et al.*, 1978). Another route has been local immunization with whole cells (Taubman and Smith, 1974; Michalek *et al.*, 1976b; Smith and Taubman, 1976) or purified antigens (Taubman and Smith, 1977). In most cases, immunization has proved effective in producing secretory IgA against *S. mutans*. Although not all workers have reported success in reducing caries (Guggenheim *et al.*, 1970; Tanzer *et al.*, 1973), the weight of evidence is that the pathogenic potential of *S. mutans* in rats can be reduced by prior vaccination. The mechanism of this protection is not clear, but some studies have indicated a reduction in the numbers of *S. mutans* associated with the animals (Taubman and Smith, 1977; Krasse and Jordan, 1977), and this would suggest a reduction in colonization.

Monkeys have also been used as experimental animals for vaccination studies with *S. mutans* as antigen. A preliminary study was undertaken by Bowen (1969) using *Macaca irus*. Since that time, the studies have been continued using this species (Bowen *et al.*, 1975), *Macaca fascicularis* (Emmings *et al.*, 1975; Evans *et al.*, 1975) and the rhesus monkey (*Macaca mulatta*) (Lehner *et al.*, 1975a, b). As with the rodent experiments, different antigens and immunization schedules have been used with varying degrees of success. However, several workers have shown that monkeys immunized by various routes developed serum and salivary antibodies which appeared to reduce the number of *S. mutans* colonizing the teeth (Bowen *et al.*, 1975; Evans *et al.*, 1975), saliva, and crevicular fluid (Caldwell *et al.*, 1977).

Thus, it has been shown in experimental animals that, at least in the case of *S. mutans*, some reduction in colonization can be induced by stimulation of a humoral response. However, the mechanisms are complex (Lehner *et al.*, 1976a; Lehner, 1977b) and involve a range of immunological functions. Several intriguing questions arise from this observation: (a) Is it possible to control all members of the oral community in this fashion or is *S. mutans* susceptible because it is an allochthonous species; (b) how extensive is stimulation of oral organisms via the gut; (c) are there oral organisms which are not recognized by the host due to shared antigens; (d) if antigenic variations occur as a mechanism for survival in the mouth, will *S. mutans* eventually "break through" the immunization; and (e) if ingested oral bacteria are coated with saliva, are they recognized as foreign in the gut? A considerable amount of work is necessary before the results with *S. mutans* can be converted into a general understanding of the relationship of the oral bacterial communities to the humoral immune system.

Studies of the lymphocytes from patients with periodontal disease have shown that antigens from oral bacteria are capable of causing transformation (Ivanyi and Lehner, 1970; Baker *et al.*, 1976; Lehner *et al.*, 1976b; Lang and Smith, 1977; Smith and Lang, 1977; Johnson *et al.*, 1978). There is good evidence that the cell-mediated immune systems recognize oral bacterial antigens (Burckhardt, 1978). The effect of stimulated cells on the bacterial communities has not been studied because, in general, the interest has centered on the tissue-destructive activities of the cell products (Lehner *et al.*, 1976b; Lehner, 1977a; Nisengard, 1977). Cells that can be detected in the gingival fluid and have direct effects upon bacteria include the polymorphonuclear and mononuclear leucocytes (Schroeder, 1977; Renggli, 1977). Ingested bacteria can be seen in polymorphs associated with the gingival area (Schroeder, 1977), although such cells may be deficient in some receptors when compared to polymorphonuclear leucocytes of the blood (Renggli, 1977; Wilton, 1977). The bacteria occupying areas close to the gingiva could, therefore, be subjected in some part to the activities of the cell-mediated immune system. Whether these activities are related solely to circulatory sensitized lymphocytes or if some local production (Waldman and Gangully, 1975) of sensitized cells occurs is not known. Certainly, most of the studies of sensitized lymphocytes have been made with cells from the blood. As with the humoral responses, the tonsillar tissue may play a role in presenting bacterial antigens to a local cell-mediated immune system (Hurtado *et al.*, 1975).

Thus, both arms of the immune system can be shown to recognize and respond to oral bacteria. To what extent these systems play a role in the control of the indigenous populations is not clear, but the observations with *S. mutans* show that control is possible. It could be that the indigenous flora in the mouth exists in that habitat because it is less susceptible to, or avoids, immune control. The effectiveness of the oral immune mechanism may have to be measured by noting, for example, the absence of organisms within the Enterobacteriaceae and *Clostridium* from oral communities.

2.5.4. Caries and Periodontal Disease

2.5.4a. Caries. Since the first observation of a relationship between bacteria and caries (Miller, 1889), there have been attempts to explain the etiology of the disease. Early studies centered on the lactobacilli as causative agents (Enright *et al.*, 1932). These bacteria were shown to be both aciduric and acidogenic and were considered to be potential pathogens. In recent years, however, the focus of research has been *S. mutans*, a species proposed as the pathogen in caries by Clarke (1924). Several recent reviews have covered the microbiology of caries, and it is unnecessary to elaborate on all the aspects of the microflora in this section (Gibbons and van Houte, 1975a; Stiles *et al.*, 1976).

The complexity of the community associated with the teeth has made it difficult to implicate a single species as the cause of caries. Certainly, the data from studies using experimental animals and some examinations of human subjects have shown that *S. mutans* must be considered as one of the main candidates for the initiation of lesions in enamel (Gibbons and van Houte, 1975b; Loesche *et al.*, 1975a; Stiles *et al.*, 1976). However, it seems unreasonable to suggest that a single organism in an ecosystem is the sole cause of enamel dissolution (Bowden *et al.*, 1976). This does not mean that all the members of a community are pathogenic [i.e., nonspecific plaque hypothesis (NSPH) (Loesche, 1976a)], as the involvement of several organisms would support the specific plaque hypothesis (SPH) by proposing that the activities of specific ecosystems produce carious lesions.

Caries can be conveniently divided into two processes when the microbial etiology of the disease is considered: (a) initiation of the lesion and (b) extension of the lesion. In the second case, there is good evidence that the ecology of the habitat produces an increase in the population of aciduric bacteria. Edwardsson (1974) made a detailed study of the advancing front of the carious lesion. He isolated a mixed flora in which gram-positive rods predominated. This group included *Propionibacterium, Actinomyces, Bifidobacterium, Eubacterium,* and *Lactobacillus. Lactobacillus* species (predominantly *L. acidophilus* and *L. casei*) were isolated most often, being detected in 48% of the 46 teeth studied. This high incidence of lactobacilli in carious dentine agrees with the results obtained by Loeche and Syed (1973), who also found *L. casei* to be the predominant species in the area.

Lactobacilli do, therefore, occupy a niche in the ecosystem in carious dentine. Although they are not the only members of the community, they occur in relatively high numbers, and as the pH of carious dentine can be low (Dirksen *et al.*, 1963), their aciduric characteristics could provide an ecological advantage. Some differences between the microflora of carious dentine associated with enamel lesions and that occurring in caries of the root surface have been noted by Sumney and Jordan (1974). These authors suggested that the site of initiation of the lesion might influence the community in carious dentine.

In contrast to their involvement in carious dentine, lactobacilli are not thought to play a significant role in the initiation of lesions in the tooth enamel. The numbers of these organisms in plaque are low (Bowden *et al.*, 1975), and they do not seem to have the ability to increase their populations on enamel surfaces in a manner similar to *S. mutans* (see subsection on Smooth Surfaces in Section 2.4.1a) (Gibbons and van Houte, 1975b; Bowden *et al.*, 1975, 1976; Marsh *et al.*, 1978). In order to cause destruction of enamel, the bacterial community must produce sufficient quantities of acid to cause the pH of the environment to drop to near 5.0 (see Section 4.2.1). Several studies have shown that *S. mutans* is associated with plaque on areas of enamel which subsequently develop a carious lesion (Ikeda *et al.*, 1973; Loesche *et al.*, 1975a; Bowden *et al.*, 1975, 1976; Hardie *et al.*, 1977). However, *S. mutans* is not obligatory for the production of lesions of smooth surface enamel (Bowden *et al.*, 1976; Hardie *et al.*, 1977). It would seem possible that the localized increases in the populations of *S. mutans* might occur in one of those ecosystems which will initiate lesion production. Longitudinal studies in Holland (Huis int'Veld *et al.*, 1978) and England (Bowden *et al.*, 1976; Marsh *et al.*, 1978) have suggested that high levels of *S. mutans* persisting over periods of time (less than 10 months) are associated with lesions that show radiographic evidence of progression. Lesions that do not appear to progress harbor fewer *S. mutans* and may be associated with a microbial community which would normally produce low or, perhaps, intermittently high levels of acid. These ecosystems may include communities which are associated with chronic lesions, and such lesions may be capable of remineralization (Silverstone, 1977). Consideration should also be given to the ability of the community to utilize lactic acid (see Section 4.4), which may decrease the rate of demineralization.

In the initiation of caries of smooth surface enamel, we have the situation where members of the indigenous flora (subsection on Smooth Surfaces in Section 2.4.1a) assume a pathogenic role, and a variation in the proportions of the populations within the microbial communities may be the only indication of pathogenic potential (Loesche, 1976a). Changes in numbers of *S. mutans* may be the only obvious and readily detectable variation which is associated with early lesions. However, communities associated with chronic lesions may have specific characteristics, one of which may be a degree of resistance to the effects of fluoride. For example, *Actinomyces* can form a high percentage of approximal plaque, and these organisms are resistant to fluoride when compared to *S. mutans* (see Section 4.7).

Caries of the fissures seems to be closely related to *S. mutans*, but as this site is a common habitat for this organism (see subsection on Occlusal Surfaces in Section 2.4.1a), its role in caries may be difficult to demonstrate. However, a recent study by Loesche *et al.* (1978) suggests that *S. mutans* increased in numbers in fissures diagnosed as having an early carious lesion. As the fissure flora is relatively simple (see subsection on Occlusal Surfaces in Section 2.4.1a), the "pathogenic" communities may be more readily detectable than those on smooth surfaces.

Caries can occur in the root surfaces of the teeth of adults (Jordan and

Sumney, 1973), and the etiology of this disease may vary in some aspects from that of enamel caries (Sumney and Jordan, 1974). It seems possible that *Actinomyces* species and other pleomorphic gram-positive rods are associated with the plaque initiating the lesion. A subsequent study by Syed *et al.* (1975) separated the plaque over the lesions into two types: In group 1, *S. mutans* made up 30% of the total cultivable flora, and in group 2, *S. mutans* was absent and *S. sanguis* made up 48% of the total flora. However, *A. viscosus* was described as the dominant organism in both groups, and as this organism can produce root caries in animals, it must be considered to have a possible role as a pathogen in root caries in humans (Jordan *et al.*, 1972).

When comparing the bacterial communities on caries-free tooth surfaces to those associated with initial carious lesions, it is difficult to see many obvious changes in populations (Bowden *et al.*, 1976). A major problem is the diagnosis of the point in time at which the lesion commenced, as there are difficulties with both radiographic and clinical assessment (Marthaler and Germann, 1970). This means that a change in the enamel surface may occur and not be detected. Subsequently, as the lesion develops, the well-described increase in *Lactobacillus* species would occur and may be the only microbiological evidence of an ecological variation.

Recognition of the possibility that specific ecosystems are related to the production of lesions should encourage more detailed work on the etiology of caries. This would take into consideration the fact that specific areas of the tooth surface, such as fissures, smooth surface enamel, and root surfaces, probably have unique combinations of etiological agents associated with early carious lesions. The results of such studies may show that prominence of *S. mutans* in an ecosystem is only one ecological parameter predisposing to a carious lesion and that other ecosystems can demineralize enamel. If this is so, methods specifically active against *S. mutans* may have to be supplemented or replaced by more generalized systems in order to give complete control of caries in all subjects (see Sections 2.5.3 and 4.7).

2.5.4b. Periodontal Disease. It has been only relatively recently that the microbial communities associated with periodontal disease have been described in any detail (Socransky, 1977). This has been due to several reasons, perhaps the most significant being the recognition of the localized nature of the disease. Following from this, sampling systems were developed to remove bacteria from the diseased sites under strictly anaerobic conditions. To a lesser extent, the taxonomy of oral anaerobes was examined. As a result of the application of these concepts, specific microbial communities have been shown to be associated with the various forms of periodontal disease.

Basically, periodontal disease is inflammation of the gingiva which can take relatively mild or severe forms. Socransky (1977) has separated periodontal disease into the following groups: (a) gingivitis, (b) periodontosis, (c) destructive periodontitis, and (d) acute necrotizing ulcerative gingivitis (ANUG). At its simplest, it can be described as gingivitis, but this condition can degenerate into well-recognized tissue-destructive diseases, such as periodontosis and destructive periodontitis. One condition which could be considered apart from the others is ANUG, which is characterized by rapid ulceration of the gingiva with invasion of tissue.

An experimental gingivitis model in humans was described by Löe *et al.* in 1965. Essentially, subjects were required to undergo intensive oral hygiene until their gingivae were free of inflammation. Plaque was then allowed to accumulate until gingivitis developed. This and another early study (Theilade *et al.*, 1966) showed the involvement of bacterial plaque in the initiation of gingivitis and also provided a model system for the examination of the microbiology, immune responses, and application of prophylactic measures. Detailed studies of the microflora of plaque during the development of gingivitis have been made (Syed and Loesche, 1978; Loesche and Syed, 1978). In these studies, quantitation of the plaque microflora was made at 0, 1, 2, and 3 weeks in the absence of oral hygiene. In the initial and first-week samples, *Streptococcus* species were the major component of the plaque—62% and 43%, respectively. *Actinomyces* species made up, respectively, 8.7% and 27.3% of the plaque community at these time periods and after 3 weeks rose to 42.2%. *A. israelii* showed a considerable increase, from 3.8% of the initial sample to 23.7% after 3 weeks. During the second and third weeks, some increases in gram-negative anaerobes were evident, including *Campylobacter sputorum*. Numbers of *Veillonella* species remained high throughout the observed time periods. Consideration of the results in relation to the plaque and gingivitis score (Loesche and Syed, 1978) indicated that *A. israelii* was associated with nonbleeding gingivitis, whereas *A. viscosus* and *B. melaninogenicus* increased with bleeding gingivitis. The presence of a variety of gram-negative anaerobes in inflamed gingival crevices has been confirmed by van Palenstein Helderman (1975). The organisms included *Campylobacter, Fusobacterium, Bacteroides,* and *Selenomonas,* both *Fusobacterium* and *Campylobacter* increasing in association with inflamed gingiva. The most recent study of the organisms associated with gingivitis, that of Slots *et al.* (1978), agrees in the main with the earlier work. Although the ratios of the populations found may vary to some extent, *Actinomyces* species were significant, and a wide range of gram-negative anaerobes was recovered. Thus, an increase in plaque bulk is associated with the development of a population of *Actinomyces*. The complexity of the community increases with time, and after about three weeks, when gingivitis and bleeding is evident, gram-negative anaerobes can contribute significantly to the community. The presence of significant numbers of gram-negative anaerobes in areas with long-standing gingivitis (van Palenstein Helderman, 1975; Slots *et al.*, 1978) has confirmed the findings of the experimental gingivitis model (Loesche and Syed, 1978; Syed and Loesche, 1978).

Periodontosis is considered to be distinct from periodontitis (Newman and Socransky, 1977) because of the age range of the patients, and is sometimes called "juvenile periodontitis" (Slots, 1976). Studies on the microbiology of this disease have used sophisticated sampling devices (Newman and Socransky, 1977) or minor surgery (Slots, 1976). In all cases, care has been taken to reduce to a minimum the exposure of the samples to O_2. Newman *et al.* (1973, 1976) were among the first to describe a particular bacterial community associated with periodontosis. When compared to normal sites, it was found that those with periodontosis had a higher number of gram-negative anaerobes, including a high percentage of organisms which could not be identified. These unidentified isolates were described as the "five periodontosis groups" and contained, (a) anaerobic saccharolytic vibrios, (b) "*Capnocytophaga*," (c) tiny gram-negative rods, (d) gram-negative saccharolytic *Bacteroides*-like organisms, and (e) gram-negative saccharolytic surface-translocating bacteria. Strains in groups (b) and (c) produced periodontal destruction when tested in gnotobiotic animals.

At the same time, Slots (1976) also described the flora associated with "juvenile periodontitis," He noticed an increase in gram-negative anaerobes and found that continued cultivation and the identification of the isolates was difficult. A further study of the flora of this type of lesion was made by Newman and Socransky (1977). A more detailed identification was made, and again organisms were isolated which did not agree with accepted descriptions. This unidentified group constituted the major component from periodontosis samples; they were gram-negative anaerobes and actively saccharolytic. Therefore, the examination of localized areas in gingival pockets associated with periodontosis revealed a flora different from that of uninvolved gingiva and included populations of anaerobes of uncertain designation.

Similar sampling systems have been applied to localized areas of the tooth and gingiva in patients suffering from destructive periodontitis. This disease is common among older people (30 years of age and up) and causes destruction of the gingiva, bone loss, and loss of teeth. It is not clear under what conditions long-standing gingivitis may develop into destructive periodontitis. However, Darwish *et al.* (1978) have examined the flora from different areas of the tooth associated with the gingiva in early periodontitis. Three areas were sampled: (a) that immediately above the gingival margin, (b) that below the gingival margin but not extending to the base of the gingival pocket, and (c) that at the base of the gingival pocket. Some variations were noted among samples from the four subjects, and, in addition, when the communities of the three areas of the tooth were examined, the supragingival showed differences from the subgingival samples in two patients. However, the variation was not consistent in these subjects. For example, Subject 1 showed no *B. melaninogenicus* subgingivally, whereas the organism constituted 21.9% and 14.8% of the cultivable flora in zones (b) and (c), respectively, in Subject 2. In contrast, in Subject 2, *A. israelii* constituted a major component of the subgingival flora, but this species was absent in Subject 1. It may have been significant that the two other subjects who gave similar results from the supragingival and subgingival samples had recently received some oral-hygiene treatment. One aspect which was common to all subjects was that the supragingival samples gave higher numbers of bacteria than the subgingival, indicating a greater density of bacteria above the gingival margin. This may be a result of the sampling techniques, which, as pointed out by Darwish *et al.* (1978), could have been relatively inefficient in recovering strict anaerobes.

In a general sense, these results are similar to those described for gingivitis by Loesche and Syed (1978) and Slots *et al.* (1978), who isolated high proportions of gram-positive organisms and also *B. melaninogenicus* from communities associated with gingivitis. Darwish *et al.* (1978) and Loesche and Syed (1978) do not make it clear which subspecies of *B. melaninogenicus* was isolated in their studies, while Slots *et al.* (1978) report that the subspecies was *B. melaninogenicus* subsp. *intermedius*.

The studies of the flora of gingival pockets associated with destructive periodontitis suggest that a unique community is present (Socransky, 1977). However, Williams *et al.* (1976) found no consistent differences between normal control sites and periodontal lesions at the level of identification that they employed. They recognize that this may have been due in part to the sampling technique, which could have removed some supragingival plaque. Despite this, these authors did isolate high numbers of gram-negative anaerobes, including *B. melaninogenicus*, from the samples. The conclusion drawn from this study was that there are no obvious etiological associations of specific bacteria with periodontitis. This is in contrast to the findings for periodontosis (Newman *et al.*, 1976; Newman and Socransky, 1977). Slots (1977b) found high levels of gram-negative anaerobes in destructive periodontitis and compared this to his findings for healthy gingivae (Slots, 1977a). The sample from periodontitis harbored fewer gram-positive organisms and more gram-negative anaerobic organisms than did the normal sites. As with the other studies, variation was detected in the bacterial populations within communities associated with periodontitis from different subjects. It seems likely that the gram-negative anaerobic flora associated with periodontitis does have some unique characteristics. Two types of communities have been proposed (Crawford *et al.*, 1975; Socransky, 1977), both including high numbers of proteolytic anaerobes. This is in contrast to periodontosis, where the flora is saccharolytic. One of the proposed communities included high numbers of *B. melaninogenicus* subsp. *asaccharolyticus*; the other contained anaerobic vibrios and the corroding organisms *Bacteroides* and *Eikenella corrodens*. Further studies will enable more definite conclusions to be reached on the numbers and types of ecosystems associated with periodontitis, but the most important advance, that of localized sampling and careful anaerobic technique, has already been made.

There has been little recent work on the bacteriology of ANUG. In his recent review, Socransky (1977) points out that spirochetes are a constant feature of the lesions, and Loesche (1976b) has discussed both the role of these organisms in ANUG and other aspects of the disease. However, in one sense, the information available on the bacteriology of ANUG is as detailed as that for the other periodontal conditions, namely, that specific communities of bacteria can be seen to be associated with pathological lesions in the gingiva. Whether these communities are the direct cause or the result of the lesion remains to be shown.

3. Formation of Dental Plaque

3.1. Introduction

In many natural habitats with low nutrient status, organisms are thought to localize on solid surfaces, as these surfaces tend to concentrate nutrients (Marshall, 1976). Although there are variations in the availability of nutrients in the mouth, it cannot be considered to have as low a nutrient status as other environments, such as the sea. Compared to this type of natural environment, the mouth is relatively rich in nutrients, which can reach high concentrations during dietary intake. Another reason for growth on surfaces is to allow the populations in the community to remain in a favorable environment. It seems possible that this may be the prime reason for organisms to be associated with surfaces in the mouth. Retention on surfaces is important if the bacteria are to survive the very efficient removal forces of saliva (Section 2.5.1). Many oral organisms have developed the capacity to adhere to surfaces while others interact with them to establish a more complex community. The result of this interaction is a surface community with some degree of organization (Section 2.4.1a). The importance of adherence in oral ecology has been reviewed adequately on several occasions during the last six years by Gibbons and van Houte (1973, 1975a) and by others (Koch, 1977); therefore, this subject will be discussed only briefly here.

3.2. Deposition of Films

Formation of bacterial films in the mouth follows a similar sequence to that in other environments and takes place in a number of stages. The first stage (a) is the physicochemical sorption of polymers to the surface. The second stage (b) is the reversible adherence of bacteria. In the third stage (c), the adherence of bacteria becomes irreversible, and the community develops by growth of the adherent bacteria and secondary colonization by other microorganisms.

a. In common with most surfaces exposed to environments containing a fluid phase, a cleaned tooth surface rapidly acquires a layer of sorbed polymers (Sonju and Glantz, 1975; Mitchell, 1976; Baier, 1977). These polymers are derived predominantly from saliva and include salivary glycoproteins and blood-group-active material (Gibbons and Qureshi, 1976). In addition to salivary polymers, it is possible that bacterial products from other habitats in the mouth bind to the tooth surface. These could include teichoic acids (Wicken and Knox, 1975) or carbohydrate polymers (Rolla, 1976; van Houte, 1976; Newbrun, 1976). Because the deposition of polymers onto the tooth is a very rapid process (Rolla, 1976), bacteria approaching the surface probably interact with the polymer layer rather than with free hydroxyapatite. The presence of saliva in *"in vitro"* systems is known to modify the adherence of bacteria to surfaces (Gibbons and Qureshi, 1976; Clark and Gibbons, 1977).

b. Saliva is most likely to be the source of the majority of the microorganisms coming close to and in contact with the tooth. These organisms may be coated with salivary glycoprotein, which in itself can have some affinity for the hydroxyapatite (Rolla, 1976) (Section 2.5.1). As the organism approaches the surface, several forces interact to determine whether or not it localizes there; these include polymer and calcium bridging (Rolla, 1976; Rutter and Abbott, 1978; Rutter, 1979).

Two factors influence the adherence of an organism to a surface: the frequency to contact and the affinity of the cell surface for the tooth surface. The number of the organisms in saliva is a major factor in determining the frequency of contact between certain bacterial species and the enamel surface. However, the tendency of the organism to be retained or removed from the surface after the initial contact depends largely on its surface characteristics and environmental factors (van Houte, 1976; Rutter and Abbott, 1978; Rutter, 1979).

In marine environments, chemoreceptors play an important role in aiding localization of motile bacteria (Mitchell, 1976; Chet and Mitchell, 1976). It has been suggested that motile bacteria can compete effectively with nonmotile strains in colonizing the gingival crevice in the mouth (Gibbons and van Houte, 1975a). Recently, the concept of colonization by motile oral bacteria has been extended by the proposition that nonmotile pathogens are carried into the gingival crevice "piggyback" on motile bacteria (Socransky *et al.*, 1978; To *et al.*, 1978). The physicochemical forces which act on bacteria as they approach a surface are balanced between van der Waals adhesive forces and the forces of electrostatic repulsion. These forces may be modified by the environment or the nature of the surface of the cells (Friberg, 1977; Rutter and Abbott, 1978).

c. The differences noted *"in vivo"* for the localization of bacterial cells in the oral cavity (Gibbons and van Houte, 1973) suggest that certain of the cells of specific bacterial species differ in their abilities to anchor to the various oral surfaces. This means that they must possess cell-surface characteristics which are unique and allow a strong interaction with a specific oral surface. These interactions are responsible for the irreversible stage of the deposition of bacterial films.

It has been proposed that polymer bridging is an important mechanism in this stage (Rutter and Abbott, 1978). Rolla (1976) has described some types of bridging which could localize the bacteria at the tooth surface. Bacterial cell surfaces are related to the composition of the cell wall, but not all wall components are exposed on the surface (Shockman *et al.*, 1976). Those cell-surface components which have been studied the most in relation to the buildup of bacterial films on teeth are polysaccharides. However, components sensitive to proteolytic enzymes also play a role, as trypsin treatment can modify the adherence of *S. salivarius* (Rutter and Abbott, 1978). Because all bacteria are likely to have some specific components associated with their surfaces, many polymer-bridging capacities exist. The emphasis given to polysaccharide receptors derives in part from the ease with which the polymers can be demonstrated on some oral bacteria. This is particularly true of the relationship between sucrose and the adherence and retention of oral streptococci on surfaces.

Various oral microorganisms are known to form glucans, fructans, and heteropolysaccharides (Guggenheim, 1970). Most attention has been drawn to the insoluble glucans synthesized from sucrose by glucosyl transferases from oral streptococci, particularly *S. mutans*. Although insoluble glucans constitute only a small proportion of plaque matrix on a weight basis (Hotz *et al.*, 1972), they do constitute important elements in the retention of organisms at the tooth surface, although the initial colonization may be sucrose independent (van Houte, 1976; Clark *et al.*, 1978). Some of the mechanisms may be similar to those proposed for reactivity of complex carbohydrates and enzymes in adherence between mammalian cells (Hughes, 1975). In particular, the Roseman hypothesis (Roseman, 1970; Roth *et al.*, 1971) is attractive, in that it involves the reaction of glycosyl transferases with glycoprotein or glycolipid receptors. Because some oral streptococci produce glucosyl transferases (Newbrun, 1976), these might be expected to interact with a glycoprotein pellicle covering the tooth.

Both teichoic and lipoteichoic acids are produced by many gram-positive bacteria (Wicken and Knox, 1975), including those of oral origin. These materials are found in association with bacterial cell walls but are also liberated into the culture fluid (Shockman *et al.*, 1976). Lipoteichoic acids are polyglycerol phosphates with glycolipid ends, and these polymers are sometimes substituted with sugars on ester-bound alanine. Lipoteichoic acids, because of their amphoteric nature and high negative charge, have been proposed as important molecules in promoting adhesion in oral bacteria. Contamination of glucans by teichoic acids may explain in part the high affinity of glucans for hydroxyapatite shown in some studies (Rolla, 1976). Because the high negative charge on teichoic acids could promote their calcium or magnesium bridging to another negatively charged surface, it is significant that chelating agents such as EDTA can cause some dissolution of dental plaque (Rolla, 1976).

Apart from the detailed studies on glucans and teichoic acids mentioned above, there is some evidence for the involvement of other surface components. Several oral bacteria show hemagglutinating activity and can carry ABO blood-group determinants (Gibbons and Qureshi, 1976; Rolla, 1976). Tooth pellicle is also known to contain active blood-group substances (Rolla, 1976), and interaction of these determinants in salivary components, pellicle, and bacteria could promote or prevent adhesion, depending on the sequence of reactions. Salivary components may reduce the adhesion of organisms by competitive inhibition for surfaces; however, organisms with exposed receptors may react with absorbed pellicle at the tooth surface.

Extension of the community beyond the populations present as a result of initial localization and retention depends on two factors: (a) reaction between the surfaces of the adsorbed bacteria and the surfaces of secondary colonizing bacteria which may have no affinity for pellicle-coated enamel and (b) growth of populations which find themselves in a suitable environment. It has been demonstrated on several occasions that different species of bacteria will aggregate with one another (Gibbons and Nygaard, 1970) (see Section 5).

This type of interaction will result in a variety of organisms adhering to the initially colonizing bacteria. The detection of these organisms on the tooth surface will depend on the growth of the population. Population growth will be governed by the parameters of the niche occupied by the organisms and by the degree of competition between these and other bacteria in the community (see Section 5). Examinations of dental plaque have shown that some oral bacterial species vary in their ability to colonize enamel; e.g., *S. sanguis* colonizes more readily than *S. mutans* (van Houte, 1976). It seems likely that certain species are able to "invade" and establish themselves within the initial community. Some studies on this type of ecological situation have been made by workers at Nijmegen using rats (Section 5).

Little is known of the rate of growth of the bacterial populations in plaque. Some estimates of the rate have been made (Gibbons, 1964a; Socransky *et al.*, 1977; Mikx and Svanberg, 1978), with mean generation times (MGTs) ranging from 3–12 hr. Plaque growth is thought to be limited by the cleansing activity of saliva. However, it seems unlikely that all of the populations in the community will grow either at the same rate or at the same time. Populations may increase either by short bursts of rapid growth followed by a relatively inactive phase or by steady, continuous slow division. Indeed, a population may exhibit a mixture of the above types of growth depending upon the environment at any given time. It would be difficult, therefore, to calculate any steady growth rate as such for dental plaque.

4. Metabolism of the Oral Microbial Community

4.1. Introduction

The vast majority of biochemical studies of the natural ecosystems of the oral cavity have been concerned with the development of dental caries. Since the caries process is dependent upon the presence of carbohydrate (Krasse, 1968), considerable emphasis has been placed on the production of acids from various sugar substrates by natural ecosystems particularly dental plaque, or by the microorganisms isolated from them. This is, of course, dependent upon the type of plaque present in the ecosystem at the time, since plaque formation per se is not dependent upon the presence of dietary sugars (Bowen, 1976). In addition, acid production by all plaque ecosystems is not the same. For example, Littleton *et al.* (1967) observed that plaque removed from patients receiving their entire diet by intubation produced very little acid when incubated with sucrose, glucose, or fructose. The vast majority of the sugar metabolized by oral microorganisms originates with the diet, although saliva does contain low levels of degradable carbohydrates. Of this dietary carbohydrate, sucrose is the most significant, since it is the main sugar of the human diet.

While it is not the purpose of this section to review extensively the influence of sugar-containing diets on the caries process, it is probably helpful to remember that the diet composition and consistency, and the frequency of eating, all have significant effects on the disease process (Krasse, 1968). The reason for this is that these dietary factors are directly related to the rate and extent of acid production by oral microbes metabolizing carbohydrates.

4.2. Exogenous Carbohydrate Metabolism

4.2.1. Dental Plaque

Subsequent to Miller's (1889) observations on the role of bacteria in dental caries, Williams (1897) and Black (1899) demonstrated that the carious lesions on the surface of the tooth enamel were always coated with adhering plaques of microorganisms. Little definitive work on the subject appeared until 1928, when Bunting *et al.* (1928) reported a positive correlation between the presence of lactobacilli in plaques and saliva and the appearance of caries. This observation was confirmed by many workers over the next 20 years (see Stralfors, 1950).

From the work of Dobbs (1932) and later Miller and Muntz (1939), lactic acid production by bacteria was thought to be a critical factor in enamel decalcification. The latter workers showed that the metabolism of glucose, sucrose, and maltose by plaque material scraped from the teeth resulted in a rapid drop in pH with the concomitant production of lactic acid; less acid was produced from lactose and starch. Muntz (1943) subsequently showed that, while lactic acid production from glucose by plaque suspensions was rapid, other acids were also formed during metabolism. He showed that the volatile acids, acetic and formic acids as well as propionic acid, were formed along with unidentified nonvolatile acids.

Using an antimony electrode placed directly in the mouth, Stephan (1940) was able to demonstrate that a rapid decrease in pH could be detected in plaque after rinsing the mouth with glucose, sucrose, or starch. Further studies (Stephan, 1944) showed that plaques capable of the lowest pH minimum were associated with individuals with the highest incidence of caries. A characteristic pH profile or "Stephan curve" is produced during plaque metabolism and shows a rapid initial pH fall and a pH minimum followed by a slower phase of pH rise. Generally, the "fasting or resting" pH obtained six or more hours after eating is 7-8, and the pH minimum observed after the addition of the various sugars was 4.7-5.2, with the pH returning to fasting levels over the next two or more hours. From this and further work (Stephan and Hemmens, 1947), it was postulated that the rapid initial fall was due to a rate of glycolysis by the plaque microflora which exceeded the rate of acid removal; the phase of pH rise was the reverse. Furthermore, it was apparent that a certain "critical" pH level must be reached before caries could occur (Stephan, 1944). In similar experiments, Stralfors (1948), while investigating the pH fall in plaques of 110 subjects receiving a glucose rinse, demonstrated a significant correlation between the number of lactobacilli and the extent of the pH minimum. In later work, Stralfors (1950) extended the work of Stephan and Hemmens (1947) and proposed the "Acid Production Diffusion Theory" to explain in greater detail the ability of plaque to maintain a pH lower than saliva. By considering the average microbial population in plaque and the buffering capacity of both plaque and saliva, it was possible to show that the observed plaque pH was regulated by the rate of acid production and the rate of diffusion from the inner plaque layer into saliva.

Studies by Kleinberg (1961) extended this work by examining the effect of the concentration of glucose and the duration of glucose addition on the rate and extent of the pH fall on plaque *in situ*. The continuous addition of glucose over a period of time caused a lowering of the steady-state pH level to a level dependent on the glucose concentration. The maximum decrease was obtained at 5% glucose, with higher levels causing a rise in the pH minimum. In experiments of the same type, Kleinberg (1967a) demonstrated that the microflora present in saliva, when incubated with glucose, produced pH profiles similar to those produced by plaque. Exact duplication of the pH–glucose-concentration curves by this salivary sediment system was dependent upon the presence of high concentrations of cells.

Obviously, the type of carbohydrate substrate added to the plaque ecosystem will have a significant influence on the pH profiles observed *in situ*. Several workers (Neff, 1967; Frostell, 1973) have demonstrated that glucose, fructose, maltose, and sucrose in food, drinks, and mouth rinses all caused similar pH decreases, since they diffuse readily into plaque. Lactose also caused a significant pH fall, but to a higher minimum than these sugars, while the pH fall with starch was relatively slow. On the other hand, the addition of sugar alcohols such as mannitol, sorbitol, and xylitol did not produce significant pH decreases in dental plaques even after prolonged exposure (Makinen and Scheinin, 1972). This has been amply and exquisitely demonstrated *in situ* through experiments employing radiotelemetric data obtained from pH electrodes built into partial dentures (Mühlemann and de Boever, 1970; Imfeld, 1977). By employing the latter technique, typical Stephan curves can be obtained with a variety of dietary carbohydrates under physiological conditions of eating, drinking, and sleeping without disturbing the structure and composition of plaque.

These studies which employed pH as the master variable, while clearly demonstrating the acidogenic nature of microbial populations in the oral cavity, nevertheless, gave very little precise information as to the metabolic properties of small oral ecosystems. A partial reason for this stems from the fact that it was not until the early 1970s that plaque was generally regarded as being heterogeneous or composed of different small microbial ecosystems (see Section 2.1). Prior to this, plaque was considered to be nonspecific, since it has not been possible to establish unequivocally the presence of overt pathogens in plaque microflora (Loesch, 1976a). With the realization that a significant portion of the plaque microbiota was obligately anaerobic and along with a subsequent improvement in the techniques for sampling and isolation through the use of reduced transport media and anaerobic chambers, it became apparent that the surfaces of the oral cavity harbor a multitude of microecosystems, each with a characteristic microbial flora (Gibbons and van Houte, 1973). This research was greatly enhanced by interest in establishing the etiological role of members of the genus *Streptococcus*, particularly *S. mutans* (Edwardsson, 1968), in the process of dental caries, since methods for the selective identification of this genus and other important oral genera were improved as a result. The local or specific nature of the microflora in dental plaque was probably no better illustrated than in studies demonstrating the distribution of *S. mutans* in the initial caries lesion or "white spot" (van Houte, 1976). It could be shown that multiple small samples on the plaque over the white spot had significantly higher numbers of *S. mutans* than samples taken from the immediate surrounding sound surface area.

Although earlier work (Muntz, 1943) had shown that lactate was not the sole metabolic end product of plaque carbohydrate metabolism, the idea prevailed until relatively recent times that lactate was the sole acid produced by plaque *in situ* (Moore, *et al.*, 1956). However, work in the last ten years has demonstrated that volatile acids are, in fact, produced during plaque metabolism *in vitro* (Gilmour and Poole, 1967; Geddes, 1972) as well as in "materia alba" from freshly extracted teeth (Tyler, 1971). For example, Gilmour and Poole (1967) observed the formation of acetate, propionate, and lactate during plaque metabolism and showed that the observed plaque pH was not related to the concentration of any one of these acids. Furthermore, the concentration of lactate was similar to, or less than, that of either acetate or propionate.

Geddes (1975) has clarified the situation with a recent study that examined the effect of sugar addition to plaque *in vivo* on the pH of plaque and the formation of lactate and volatile acids. She showed that all samples contained acetate, propionate, *n*-butyrate, and L(+)- and D(-)-lactate. The volatile acids were high in fasting plaque, while lactate isomers were increased significantly (5- to 8-fold) upon the addition of one lump of sugar. The addition of multiple sugar lumps resulted in a slower pH recovery and an increased level of L(+)-lactate 30 min after the addition of the sugar. It could be shown that the pH minimum of the Stephan curve was due to the production of L(+)- and, to a smaller degree, D(-)-lactate, while at near-neutral pH values the volatile acids were dominant. Very little difference was seen between the total individual acids produced from glucose or sucrose.

Recent work by Gilmour and co-workers (1976) has not only demonstrated the heterofermentative nature of plaque metabolism but has also shown that the assumption that carbohydrate metabolism is similar in various plaque ecosystems is a fallacy. Metabolism by small plaque samples from localized tooth sites resulted in the formation of acetate, lactate, and formate as major products, with propionate and butyrate as minor products; however, the anionic profiles showed significant differences between different ecosystems. Clearly, much more work is required on the metabolic properties of very small plaque samples from localized areas in order to associate metabolic activity with microbial components as a means of understanding the etiology of dental caries.

Much less is known concerning the metabolism of both supra- and subgingival plaque associated with periodontal disease, since relatively little is known concerning the etiological agent(s) for the disease (Socransky, 1977). It has been assumed, however, that microbial enzymes, metabolic end products, and cellular constituents are involved in the initial phases of the disease, which leads to the destruction of the periodontium and, eventually, the alveolar bone in which the tooth structure is embedded (Socransky, 1970).

4.2.2. Pure-Culture Studies

4.2.2a. Streptococci. In the absence of sufficient data from *in vivo* oral ecosystems, many assumptions concerning the metabolic properties of these systems have come from the study of the biochemical properties of species isolated in pure culture from the microbial community. It is perhaps natural that the most attention in this respect has been given to organisms thought to be involved in the etiology of the two major oral diseases: dental caries and periodontal disease. Although the etiology of both diseases is unknown, much more is known about dental caries than periodontal disease because of the involvement of microbial acid production in tooth destruction. Consequently, bacteriologists have, for many decades, been studying the properties of the predominant acid-producing microorganisms found in dental plaque and saliva. Two genera which have been singled out for special attention at different periods throughout the history of dental research are *Lactobacillus* and *Streptococcus.*

The lactobacilli were popular for almost the first half of the twentieth century since Kligler (1915) demonstrated that the lactobacilli, being both acidogenic and aciduric, were found in high numbers in carious lesions. This stance was augmented by other workers, notably Enright *et al.* (1932), who, in a longitudinal study, showed the appearance of lactobacilli in saliva preceding the appearance of carious lesions. In addition, the individuals with high *Lactobacillus* counts had caries more frequently than individuals with lower counts. However, in a subsequent detailed study, Stralfors (1950) was able to show that the streptococci were present in dental plaque in far higher numbers (i.e., 10^{-4}-fold) than the lactobacilli and were also capable of rapid acid production from sugar substrates.

The emergence of the oral streptococci as important members of the plaque ecosystem and possible etiological agents for human dental caries has occurred as a result of the accumulation of various types of data: (a) Members of the genus *Streptococcus* constitute a significant fraction of the oral community (Carlsson, 1967; Loesche *et al.*, 1972; Loesch and Syed, 1973; Bowden *et al.*, 1975); (b) various species, notably *S. mutans*, produce extensive tooth decay in animals fed high-sucrose diets (Keyes, 1968; Krasse and Carlsson, 1970); and (c) *S. mutans* has been linked to tooth decay in humans (Krasse *et al.*, 1968; de Stoppelaar *et al.*, 1969; Jordan *et al.*, 1969; Gibbons *et al.*, 1974; Loesch *et al.*, 1975a). Although the results from a recent longitudinal study (Bowden *et al.*, 1976) indicate that *S. mutans* is not a mandatory requirement for the initiation of a carious lesion, the streptococci as a group are, nevertheless, important members of the oral microbial community.

In addition to their high numbers in the oral cavity, the streptococci possess biochemical properties which suggest that they play an important role in the colonization and metabolism of the various oral ecosystems, particularly dental plaque. Oral strains of the genus are capable of metabolizing a variety of sugar substrates with the production of acid end products, as well as extracellular (see Section 3.2) and intracellular polysaccharides (see Section 4.3). Since the catabolism of the latter polysaccharides will be discussed in Section 4.3, the discussion here will be restricted largely to exogenous sugar metabolism.

The homofermentative nature of the oral streptococci has been demonstrated in experiments employing both washed cells (Hamilton, 1968; Tanzer *et al.*, 1969, 1972; Yamada *et al.*, 1970) and cells growing in batch culture (Jordan, 1965; Drucker and Melville, 1968; Robrish and Krichevsky, 1972) with glucose or sucrose as substrates. Acetate and formate, as well as ethanol, were also observed in these experiments but to a lesser extent. Interestingly, no significant differences have been observed in metabolic end products between cariogenic (i.e., *S. mutans* strains) and noncariogenic species (Jordan, 1965; Drucker and Melville, 1968).

More recent results, however, suggest that washed cells and cells growing in batch culture do not accurately reflect the metabolism of the streptococci *in vivo*. Analysis of the metabolic products of dental plaques of germ-free rats monoassociated with *S. mutans* showed that acetate and ethanol were present in addition to lactic acid (van der Hoeven, 1976). Since strains of *S. mutans* and *S. sanguis*, and to a lesser extent *S. salivarius* and *S. bovis*, are heterofermentative under conditions of carbon, but not nitrogen, limitation in continuous culture (Carlsson and Griffith, 1974), it would appear that the carbohydrate source rather than the nitrogen source is limiting growth of *S. mutans* and probably other acidogenic microbes in plaque. This is contrary to the earlier observations of Hotz *et al.* (1972) and Carlsson and Johansson (1973), who suggested that the growth of the plaque flora is not limited by carbohydrate.

A further factor generally not taken into consideration is the effect of growth rate on carbohydrate metabolism by the oral streptococci. Studies with *S. mutans* grown in continuous culture under conditions of carbon limitation have demonstrated that as the growth rate approaches the maximum (μ_{max}), with an MGT of 1.38 hr, metabolism progressively changes from hetero- to homo-fermentation (Ellwood *et al.*, 1974; Mikx and van der Hoeven, 1975; Ellwood, 1976). Since batch-grown cells grow in the logarithmic phase at or near μ_{max} (Ellwood *et al.*, 1974), it is not surprising that many workers employing this technique report streptococci as producing only lactate during glucose or sucrose metabolism. Since mature dental plaque is thought to undergo only two to three cell divisions per day (i.e., MGT of 8–12 hr) (Gibbons, 1964a) and is carbon limited (van der Hoeven, 1976), it is apparent that *S. mutans* and other oral streptococci probably metabolize sugars *in vivo* with the production of acetate, formate, and ethanol, as well as minor amounts of lactate (Carlsson and Griffith, 1974; Yamada and Carlsson, 1975). Only under the transitory conditions of dietary sugar intake would dental plaque possess an excess of carbon. In this situation, the streptococci would revert to homofermentation as long as the sugar concentration was in excess (Carlsson and Griffith, 1974; Hamilton and Ellwood, 1978b). Thus, one can visualize the oral streptococci alternating between hetero- and homofermentation depending on the dietary intake of sugar. This process would be in keeping with the fermentation patterns observed with plaque *in vivo* (Gilmour and Poole, 1967; Geddes, 1975), in which lactate was high immediately following sugar ingestion but declined with time. This decline of lactate was associated with an increase in volatile acids and ethanol.

The reason for the variation in fermentation patterns in the oral streptococci is associated with the regulation of both lactic dehydrogenase (LDH) and pyruvate formate lyase. Extending the earlier observations of Wolin (1964), Wittenberger and co-workers have established that LDH in *Streptococcus faecalis* (Wittenberger and Angelo, 1970) and strains of *S. mutans* (Wittenberger *et al.*, 1971; Brown and Wittenberger, 1972) has an absolute requirement for the activator fructose-1, 6-diphosphate (FDP). Yamada and Carlsson (1975) subsequently demonstrated that cells of *S. mutans* JC2, grown in continuous culture under conditions of both glucose excess (nitrogen limited) and glucose limitation, possessed similar FDP-activated LDH activity. In addition, the cellular concentration of FDP was high, while that of phosphoenolpyruvate (PEP) was low, in glucose-excess cells; in glucose-limited cells, the reverse was true. This low level of FDP in the latter cells explains why glucose metabolism by these cells resulted in the formation of acetate, formate, and ethanol but not lactate. Lactate was the main fermentation product of the cells grown with an excess of glucose. By comparison, the same study also demonstrated that *S. mutans* FIL and *S. bovis* ATCC 9809 produced mainly lactate under both growth conditions. *S. mutans* FIL had similar levels of the glycolytic intermediates but possessed a FDP-independent LDH, while the *S. bovis* enzyme was only dependent on FDP at low pyruvate concentrations. These and earlier results (Carlsson and Griffith, 1974) indicated that the activity of LDH is regulated by the cellular level of FDP, and this, in turn, influences the type of fermentation products formed.

Yamada and Carlsson (1976) have also demonstrated that the activity and regulation of pyruvate formate-lyase is an important factor in the metabolic sequence. The activity of the enzyme, which essentially converts pyruvate to acetate and formate, was much higher in *S. mutans* JC2 grown under conditions of glucose limitation than in cells grown with a nitrogen limitation. In the latter cells, the concentrations of FDP and D-glyceraldehyde-3-phosphate (GAP) were high, and since pyruvate formate-lyase is inhibited by GAP while FDP activates LDH, the sole product of metabolism is lactate (Fig. 1). Under glucose-limited

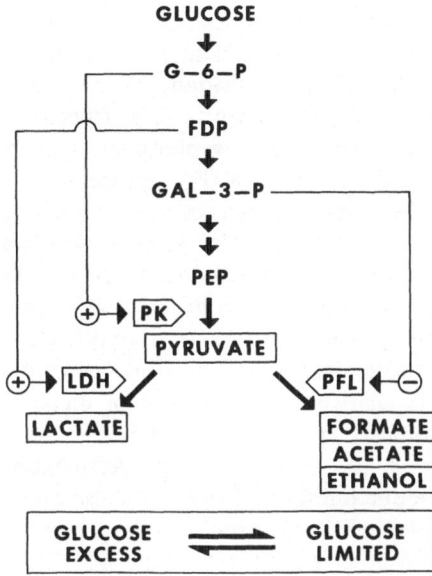

Figure 1. Pathways of carbohydrate metabolism by *Streptococcus mutans* under conditions of glucose excess and glucose limitation. G-6-P, Glucose-6-phosphate; FDP, fructose-1,6-diphosphate; GAL-3-P, glyceraldehyde-3-phosphate; PEP, phosphoenolpyruvate; PK, pyruvate kinase; LDH, lactate dehydrogenase; PFL, pyruvate formate-lyase; (+), positive effector; (−), negative effector. From Yamada and Carlsson (1976).

conditions, on the other hand, the cellular concentration of these two intermediates (FDP and GAP) is low, thereby deactivating LDH and releasing the inhibition of pyruvate formate-lyase with the resulting production of formate, acetate, and ethanol by these cells. The ability of *S. mutans* FIL to form lactate under both conditions was ascribed to a much higher level of LDH (FDP-independent) activity compared to that for pyruvate formate-lyase. It is, therefore, apparent that formate-lyase in this organism does not compete effectively with LDH for pyruvate, as the LDH K_m (pyruvate) was the same following growth under both conditions (Yamada and Carlsson, 1976).

Apart from the experiments concerned with the end products of carbohydrate metabolism by the oral streptococci, considerable interest has been shown in sugar transport by these bacteria, since effective inhibition of this process would effectively reduce dental caries. Studies on the mechanism of fluoride inhibition of sugar metabolism (see Section 4.7) by *S. salivarius* were the first to demonstrate that glucose transport in the oral streptococci occurred via the PEP phosphotransferase (PT) system (Kanapka and Hamilton, 1971). Subsequent work showed the presence of the PT system in other strains of oral streptococci, including *S. mutans* (Schachtele and Mayo, 1973), although significant differences in specific activity for the glucose-PT system have been observed between the various strains of this organism (Hamilton, 1977).

Recent continuous-culture studies with *S. mutans* Ingbritt have shed more light on the probable role of the PT system in this organism. Growth of the organism at a rate comparable to that for *in vivo* plaque (i.e, MGT of 12 hr) (Gibbons, 1964a) demonstrated that at low pH, glucose-PT was repressed (Hamilton and Ellwood, 1978). This repressive effect could also be shown following growth with excess glucose (e.g., nitrogen limitation) (Ellwood *et al.*, 1979), with limiting phosphate (Ellwood, unpublished data), and with a fast growth rate (i.e., MGT approaching 1 hr) (Hunter *et al.*, 1973; Ellwood *et al.*, 1979). Since the cells possessing repressed levels of the PT system were, nonetheless, capable of active growth, this indicates that at least one additional sugar-transport system functions in the organism. Only under conditions of slow growth (MGT of less than 6 hr) with a glucose limitation was the activity of the glucose PT system sufficient to support growth (Ellwood *et al.*, 1979). One would predict that under *in vivo* plaque conditions of carbon limitation, except during the relatively short periods of dietary sugar intake, the PT system would predominate in cells of *S. mutans*.

4.2.2b. *Other Genera.* As mentioned previously, the central role of acid production by oral microorganisms in the process of dental caries has resulted in considerable research into the properties of the acidogenic oral streptococci and, to a lesser extent, the lactobacilli. The metabolism of other members of the oral microbial community has been largely ignored.

Members of the genus *Actinomyces* are predominant members of the oral microflora but are less acidogenic than either the streptococci or the lactobacilli. The anaerobic fermentation of glucose by *A. naeslundii* in the absence of CO_2 results in the formation of only lactate, whereas with CO_2, acetate, formate, and succinate are the major products. Aerobically, CO_2 was required only to initiate growth (Pine and Howell, 1956; Buchanan and Pine, 1963). More recently, it has been shown that fermentation occurs primarily via the Embden-Meyerhof pathway with either CO_2 or O_2 being required for maximum growth (Buchanan and Pine, 1967).

Much less is known about the metabolism of other species of *Actinomyces*, although *A. viscosus* is known to produce lactate and succinate from glucose under anaerobic conditions (Howell and Jordan, 1963). Continuous-culture studies with a human strain of *A. viscosus* (GN 431/75) demonstrated that at dilution rates between $D = 0.025$ and 0.2 hr^{-1} in a defined medium under a glucose limitation, the organism produced lactate, succinate, acetate, formate, and ethanol (Ellwood *et al.*, 1977). Of these end products, only the concentration of succinate was altered to any extent by changes in the growth rate, falling from 6.5 to 2.5 mg/ml when the dilution rate was increased from $D = 0.025$ to 0.2 $hr.^{-1}$ Interestingly, during the variation in the growth rate, the culture changed from discrete cells ($D = 0.025$ hr^{-1}) with only minor branches and few agglomerates to highly branched cells which formed tenacious films on the walls of the chemostat ($D = 0.2$ hr^{-1}). The preferential growth of the organism in a film is coincident with the observation that *Actinomyces* species are components of dental plaque in a wide variety of animals (Bowden *et al.*, 1978).

Transport studies with the same human strain of *A. viscosus* (Hamilton and Ellwood, 1977) demonstrated that the organism contained low levels of the PEP PT system (2.0 nmol/mg cells per min). Furthermore, the K_m for glucose uptake by washed cells was 2×10^{-6} M, with the inherent glycolytic rate increasing from 12 to 40 nmol/mg dry weight of cells per min as the MGT was reduced from 24 to 2 hr. This metabolic activity is, however, sluggish when compared to that of *S. mutans*. A comparison of the glycolytic activity of *S. mutans* Ingbritt and *A. viscosus* GN 431/75 grown in the same defined medium with a glucose limitation at $D = 0.05$ hr^{-1} showed the former organism to be 20 times more active (e.g., 327 to 17 nmol acid/mg cells per min) than *A. viscosus* (Hamilton and Ellwood, unpublished data). A somewhat similar finding was made recently by Ellen and Onose (1978), who demonstrated that *S. mutans* colonies on tryptic soy agar accumulated more acid than colonies of *A. viscosus* and *S. mitis*. This confirms the work of others (Larje and Frostell, 1968; Charlton *et al.*, 1971).

These observations, coupled with animal studies showing that *A. viscosus* is weakly cariogenic in monoassociated gnotobiotic rats (van der Hoeven *et al.*, 1974), have suggested to some (Ellen and Onose, 1978) that *Actinomyces* species are not important in caries etiology. However, a variety of evidence suggests that these organisms, although possessing low glycolytic activity, may be more important than originally suspected. Firstly, animals monoinfected with *A. viscosus* produce high plaque scores with glucose or sucrose diets despite low caries activity (Llory *et al.*, 1971; van der Hoeven *et al.*, 1974). This suggests that, in combination with other indigenous flora under *in vivo* conditions, they would support caries activity. Such was, in fact, the case when *A. viscosus* Nyl was successfully implanted with the indigenous flora in SPF (specific-pathogen-free) rats (van der Hoeven *et al.*, 1974). Significant caries was observed with sucrose and, to a lesser extent, with glucose, illustrating the cariogenic interaction between *Actinomyces* species and the other microflora. Furthermore, various *Actinomyces* species appear to be inherently fluoride resistant (Beighton and Colman, 1976; Beighton and MacDougall, 1977), as well as possessing the ability to synthesize large quantities of glycogen in the presence of low concentrations of sugar (Hamilton and Ellwood, unpublished data). These are factors which would give members of this genus a selective advantage over other organisms in the ecosystem.

4.3. Endogenous Metabolism

The microorganisms in the various ecosystems in the oral cavity are normally exposed to an intermittent and variable supply of fermentable carbohydrate in the diet. As a consequence, many of the predominant organisms rapidly convert dietary sugars to extracellular glucans and fructans and to intracellular polysaccharides, in addition to acid end products. Evidence is now available to indicate that both the extra- and intracellular polymers are catabolized by the indigenous flora in the absence of an exogenous sugar supply, presumably to provide maintenance energy (Dawes and Ribbons, 1964).

The ability of oral microbes to undergo endogenous metabolism was first recognized by Gibbons and Socransky (1962), who observed the production of intracellular polysaccharide (IPS) by bacteria in dental plaque. These workers were able to demonstrate a correlation between the proportion of plaque microorganisms producing IPS and dental caries. This has been confirmed by other investigators (Loesch and Henry, 1967; van Houte *et al.*, 1969a). Obviously, plaque possessing a high proportion of IPS-containing microbes would have the capacity to maintain acid production for prolonged periods in the absence of an exogenous sugar source and would thereby contribute significantly to enamel decalcification. Furthermore, the ability of an organism to synthesize such an energy-storage compound should give such an organism a selective advantage in its ecosystem (van Houte and Jansen, 1970; de Stoppelaar *et al.*, 1970).

A variety of factors appears to govern the formation of IPS in plaque. For example, the fraction of the microflora in plaque capable of forming iodine-staining polysaccharide varies with the concentration of dietary carbohydrate, the proportion being directly related to the level of carbohydrate in the diet (van Houte, 1964; de Stoppelaar *et al.*, 1970). Furthermore, studies employing electron microscopy (EM) (van Houte and Saxton, 1971) have demonstrated that plaque at the plaque–tooth interface was composed of microorganisms with thick cell walls that contain large amounts of IPS. Subsequent studies employing EM autoradiography (Saxton, 1975b) demonstrated that only 16% of the organisms at the saliva interface synthesized IPS, whereas 52% of those near the tooth surface formed the polymer in the presence of glucose. Critchley (1969) has suggested that a deficiency of certain essential amino acids in an unbalanced growth situation promotes the synthesis of IPS and the thickening of cell walls.

Of the various genera capable of synthesizing IPS in the oral cavity (Berman and Gibbons, 1966), the streptococci have been studied most extensively. Early results demonstrated that "cariogenic streptococci" were strong and genetically stable IPS producers, while "noncariogenic" strains were more variable and gave rise to negative variants (Berman and Gibbons, 1966). Strains synthesizing large amounts of IPS were also observed to be more acidogenic in the presence of exogenous carbohydrates (van Houte *et al.*, 1969b).

Many of the early studies were concerned with the characteristics of IPS synthesis in *S. salivarius* (Hamilton, 1968; van Houte and Jansen, 1968a) and *S. mitis* (Gibbons and Kapsimalis, 1963; Gibbons, 1964b; Weiss *et al.*, 1965; Berman *et al.*, 1967; van Houte *et al.*, 1969b, 1970). These studies showed that IPS synthesis was maximum during the early stationary phase of growth and, in the absence of exogenous sugar, IPS was readily degraded to lactic acid. The prolonged exposure of washed cells to high levels of glucose resulted in a 50%

increase in cell weight due to polysaccharide synthesis (Gibbons, 1964b; Hamilton, 1968). Similar results have been obtained with *L. casei* (Hammond, 1971).

The intracellular polymer synthesized in *S. salivarius* and *S. mitis* possesses a structure typical of glycogen (Bramstedt and Lusty, 1968; van Houte and Jansen, 1968a; Khandelwal *et al.*, 1972) and appears in distinct cytoplasmic granules (Berman *et al.*, 1967; Hammond, 1971).

Berman and Gibbons (1966) first reported on the IPS-producing ability of "cariogenic" streptococci (*S. mutans*) and indicated that these strains consistently formed abundant IPS, while "noncariogenic" strains were more variable in IPS formation and frequently gave rise to IPS-negative variants. Interestingly, 60% of the viable organisms from carious lesions were strong IPS producers, while only 13% of those in caries-inactive plaque were strong producers. Subsequently, work by van Houte *et al.* (1970) showed that a variety of strains of *S. mutans* synthesized large amounts of IPS from glucose, sucrose, and fructose, and most were stable IPS producers. Since a few "cariogenic" strains were variable IPS producers, the authors concluded that an absolute correlation between the degree and stability of IPS synthesis and cariogenic potential could not be made. More recent work (Freedman and Tanzer, 1974; Freedman and Coykendall, 1975) has confirmed this by demonstrating that representatives of serotype IIId of *S. mutans*, although capable of producing caries in animal systems, were nevertheless weak IPS producers.

The relationship between cariogenicity and IPS production has been further examined in experiments in which SPF rats were infected with IPS-defective mutants of *S. mutans* NCTC-10449 and *S. mutans* Ingbritt-1600, both *Ic* serotypes (Tanzer *et al.*, 1976). The IPS-defective mutants colonized the tooth surface of the infected rats as readily as the wild-type parent but produced fewer carious lesions than the parental strain. Since many of the mutant strains produced higher caries scores than the uninfected controls, it was concluded that determinants, in addition to IPS-synthesizing ability, are involved in the carious process.

Although the IPS-defective mutants are known to form less polymer than their wild-type parent, the actual genetic lesion in each case has not been determined. Since the effect of the mutational event on overall glycolytic activity was not reported, it is possible that the lesion could occur either during glucose transport or in the Embden–Meyerhof pathway and still result in low IPS formation. However, such a lesion would produce a cell with drastically reduced ability to ferment sugar, with a resultant decline in acid production, which would be the cause of lower caries scores. Thus, the extent of IPS formation by the mutant strains need not have been a determining factor in the results observed.

Unlike previous work with *S. salivarius* and *S. mitis*, little definitive data have been published as to the structure of the polymer synthesized in *S. mutans* strains. Dipersio *et al.* (1974) were able to demonstrate and quantitate IPS granules in stationary-phase cells of strains FA-1 and OMZ-176 with EM following periodic acid–chlorite treatment and uranyl–lead citrate staining. Chemical analysis of the polymer in FA-1 showed that 90% was composed of glucose. Hamilton (1976), employing glucoamylase to determine the glycogen content of *S. sanguis* 10556, *S. salivarius* 25975, and six strains of *S. mutans*, demonstrated that the *S. mutans* strains formed relatively small quantities of glycogen compared to *S. salivarius* under conditions supporting "maximum" glycogen synthesis. Although the rate of glycolysis for the latter organism was only twice that of the other strains, the rate of glycogen synthesis was 5–80 times greater than *S. mutans*. Interestingly, determination of the specific activity of the glycogen-synthesizing enzymes, ADP glucose synthase and ADP glucose transferase, indicated that these enzymes were capable of supporting a greater rate of synthesis in *S. mutans* than was actually observed. It is, therefore, apparent that *S. mutans* cells synthesize less "glycogen-like" material than had been thought from studies employing iodine staining and glucose determinations following acid hydrolysis of cellular polysaccharide (Dipersio *et al.*, 1974). Since *S. mutans* Ingbritt, a serotype *c* strain, was shown to contain much higher levels of metabolizable total carbohydrate than glycogen (Hamilton and Ellwood, 1978a), the apparent discrepancy may be simply a matter of structure. For example, the cell could synthesize several carbohydrate polymers, only one of which contains the typical α-(1 → 4) and α-(1 → 6) linkages of glycogen. Some evidence exists (Ellwood and Hamilton, unpublished data) that the ethanol-precipitable material within *S. mutans* Ingbritt contains rhamnose, which is a cell-wall constituent (Hardie and Bowden, 1974b). Therefore, since cell-wall thickening occurs with *S. mutans* under conditions of amino acid deprivation (Mattingly *et al.*, 1976, 1977), it is possible that a polysaccharide cell-wall precursor is synthesized and stored in cells in a form similar to that of glycogen and is capable of being degraded for energy production under starvation conditions.

Considerable amounts of research have been directed at the relationship between the synthesis of extracellular polysaccharide (EPS) and the colonization of the tooth surface (see Section 3). Often it is forgotten that these extracellular polymers are "endogenous" energy reserves providing carbon substrates to the microflora in the intervals between dietary intake. The majority of the EPS in dental plaque is of the glucan or fructan type (Critchley *et al.*, 1967), although a variety of low-molecular-weight oligosaccharides, disaccharides, and pentoses are also found (Hotz *et al.*, 1972). Wood (1967, 1969) demonstrated that the total and extractable hexoses in dental plaque, representing both IPS and EPS, were readily metabolized upon incubation at 37°C. For example, 50% of the hexose fractions were degraded with the production of acid in a 24-hr period.

Of the two main extracellular polymers, fructans are more readily degraded by the microflora in plaque (Wood, 1964; Gibbons and Banghart, 1967; van

Houte and Jansen, 1968b; Da Costa and Gibbons, 1968; Manley and Richardson, 1968; Leach *et al.*, 1972). Fructans are synthesized in the presence of sucrose and degraded in the absence of preferred substrates, such as glucose, by a process involving the induction of fructan hydrolase (Wood, 1964; van Houte and Jansen, 1968b; Da Costa and Gibbons, 1968). Published evidence suggests that much of the fructan-hydrolyzing activity in dental plaque originates with the oral streptococci (van Houte and Jansen, 1968b; Da Costa and Gibbons, 1968), although *L. casei* and *A. viscosus* are also capable of fructan degradation (Parker and Creamer, 1971; Rosan and Hammond, 1974; Palenik and Miller, 1975). From these studies, it is apparent that organisms such as *S. salivarius* and *A. viscosus* (van der Hoeven *et al.*, 1976) are capable of both synthesizing levan and degrading it in a manner analogous to that for IPS. Fructan hydrolase in plaque streptococci was shown to be induced during growth on levan, inulin, and sucrose, but synthesis was repressed by glucose and fructose (Da Costa and Gibbons, 1968).

For some time, it was thought that glucans in the plaque matrix were relatively immune to microbial degradation (Gibbons and Banghart, 1967). Subsequently, however, it was shown that glucans were degraded by various strains of *Streptococcus* (Parker and Creamer, 1971; Walker and Pulkownik, 1973; Guggenheim and Burckhardt, 1974; Staat and Schachtele, 1974; Dewar and Walker, 1975) as well as by *Fusobacterium fusiforme* (Da Costa *et al*, 1974), *B. ochraceus*, and various strains of *Actinomyces* (Staat *et al*, 1973). Glucan hydrolases in plaque, like fructan hydrolases, appear to be induced to increased levels by sucrose-rich diets (Gawronski *et al.*, 1975). The combined action of the various hydrolases produces glucose and oligosaccharides from the degradation of matrix glucans principally of the α-(1 $\rightarrow$ 6)-linked type (Hotz *et al.*, 1972). It is known that the insoluble glucan of plaque has a high α-(1 $\rightarrow$ 3) glycosidic bond content and is resistant to degradation by the oral microflora (Dewar and Walker, 1975).

Recent research has indicated that some strains of *S. mutans* produce a cell-associated *endo*-α-1,6-glucan 6-glucanohydrolase, while others produce an *exo*-hydrolytic enzyme (Guggenheim and Burckhardt, 1974; Schachtele *et al.*, 1975; Dewar and Walker, 1975). An oral isolate of *A. israelii* produces *endo*-hydrolytic activity, while *B. ochraceus* produces both *exo*- and *endo*-enzyme activity (Staat *et al.*, 1973; Schachtele *et al.*, 1975). Although glucan hydrolase activity can be found in oral strains unable to synthesize EPS, it has, nevertheless, been suggested that the enzyme is involved in the synthesis of insoluble glucan (mutan), which is predominantly α-(1$\rightarrow$3) linked (Walker, 1972; Schachtele *et al.*, 1975). Low levels of the *endo*-hydrolase activity are thought to modify the glucan synthesized by glucosyl transferases so that the molecules contain a higher proportion of α-(1$\rightarrow$3) linkages, thus making the polymer less soluble and less susceptible to microbial degradation.

Apart from microbial polysaccharides, oral microorganisms appear to have a limited ability to degrade macromolecules formed by the host. The plaque microflora is known to cleave carbohydrate moieties from salivary glycoproteins (Leach and Hayes, 1968; Leach and Melville, 1970; Fukui *et al.*, 1973), which may provide a low-level energy source when preferred substrates are unavailable.

4.4. Acid Utilization

As mentioned previously (Section 4.1), the addition of sugar to dental plaque results in a rapid fall in pH to a minimum, which may be low enough to promote enamel decalcification and eventually result in a carious lesion. The pH minimum can be maintained by the continued metabolism of both extra- and intracellular carbohydrate, but this minimum is also influenced considerably by the rate and extent of acid utilization and base formation, as well as by saliva flow (Edgar, 1976).

A small group of organisms in dental plaque is capable of utilizing metabolic acids, in particular, lactate. These include both types of species of *Veillonella* (Rogosa, 1965), *Neisseria* species (Hoshino *et al.*, 1976), and *E. alactolyticum* (Holdeman and Moore, 1977). The degradation of lactate by *Neisseria* species and *B. matruchotii* results in the formation of acetate and CO_2 under aerobic conditions (Iwani *et al.*, 1972; Hoshina *et al.*, 1976). Since the plaque ecosystem is largely anaerobic (Ritz, 1967; Kenney and Ash, 1969), except during early plaque development, one would imagine that these organisms may not contribute greatly to plaque acid consumption.

On the other hand, the obligate anaerobes in the genus *Veillonella* are present in significant numbers in dental plaque (Loesche *et al.*, 1972; Loesche and Syed, 1973; Bowden *et al.*, 1975) and saliva (Gordon and Jong, 1968; Liljemark and Gibbons, 1971) and are capable of metabolizing a variety of short-chain acids (Whiteley and Ordal, 1957; Rogosa, 1964). Members of this genus are asaccharolytic because of lesions in the glycolytic pathway. Rogosa *et al.* (1965) reported that the *Veillonella* species lacked hexokinase, while Michaud and Delwiche (1970) reported that their sheep-rumen strain of *V. alcalescens* was devoid of hexokinase, phosphoglyceromutase, and pyruvate kinase. In contrast, an oral strain of *V. parvula* was shown to contain pyruvate kinase regulated in a manner similar to that found in glycolytic organisms (Ng and Hamilton, 1975).

Veillonella species grow well on lactate, with the lactate's carbon distributed to various cellular components (Michaud *et al.*, 1970; Ng and Hamilton, 1974) by gluconeogenesis employing one or more CO_2-fixation reactions (Ng and Hamilton, 1973). The metabolism of lactate to propionate, acetate, CO_2, and H_2 by these species (Foubert and Douglas, 1948; Ng and Hamilton, 1971) can be viewed as a secondary stage in the plaque food chain. The primary system is lactate production by the acidogenic microorganisms. Since the products of *Veillonella* metabolism are weaker acids than lactate, the rapid conversion of the latter acid in the habitat should theoretically reduce enamel decalcification. No comparison between communities with high and low populations of *Veillonella* has been carried out in relation to caries activity in humans. However, the diassociation of *V. alcalescens* with *S. mutans* in germ-free rats fed a high-sucrose diet resulted in a reduction in caries compared to those animals inoculated with only *S. mutans* C67-1 (Mikx and van der Hoeven, 1975). Mikx and van der Hoeven, in an accompanying mixed-continuous-culture study with the two organisms, demonstrated that lactate was not detectable in the culture except at high dilution rates with high glucose concentrations. In these experiments, where *V. alcalescens* represented 10–25% of the total colony-forming units, the major metabolic end products were acetate, propionate, and ethanol, indicating that the organism was part of a food chain. More recent studies with gnotobiotic rats have shown that the proportion of *V. alcalescens* in the mixture increases in combination with a high-lactate-producting strain of *S. mutans* (FIL) and decreases in relationship to a low-lactate-producing strain of *S. mutans* (C67-1) (van der Hoeven *et al.*, 1978).

4.5. Nitrogen Metabolism

The microflora in dental plaque is known to generate alkaline compounds, since fasting levels of plaque pH are frequently higher than that of saliva (Kleinberg and Jenkins, 1964). The extent to which alkali production by plaque microflora influences the caries process is unknown, but Kleinberg *et al.* (1976) have suggested that caries may be more a deficiency of plaque microbes to produce adequate base than their ability to produce acid.

Some base production in plaque occurs through the metabolism of urea in saliva to ammonia and CO_2 (Kleinberg, 1967b), while other reactions include the decarboxylation of plaque amino acids to amines (Hayes and Hyatt, 1974). These amino acids are derived from the hydrolysis of salivary proteins and glycoproteins by oral microorganisms (Leach and Melville, 1970; Cowman *et al.*, 1975).

Apart from buffering capacity, saliva also contains a component which not only stimulates glycolysis but also stimulates base formation (Kleinberg *et al.*, 1973). The concomitant function of these two saliva-induced processes can result in the rapid clearance of plaque carbohydrate. Under these conditions, the pH drop is small, and, consequently, the plaque pH rapidly returns to normal levels (i.e., pH 7.0). Such a mechanism occurs through the action of a salivary "pH-rise factor" and may operate in individuals resistant to dental caries (Kleinberg *et al.*, 1976). The "pH-rise factor," or sialin, is a basic tetrapeptide with the composition glycine–glycine–lysine–arginine, which not only promotes glycolysis but also stimulates base formation (Kleinberg *et al.*, 1976). Sialin appears to be different from the other glycolysis-enhancing fractions in saliva (Hay and Hartles, 1965; Mishiro *et al.*, 1966; Holbrook and Molan, 1975) by its ability to stimulate base formation. Sialin is active at low concentrations (40 μM), is heat stable, highly soluble in water, and is tasteless. Kleinberg and co-workers (1976) have demonstrated that a variety of arginine-containing peptides can also stimulate a pH-rise phenomenon similar to that observed with sialin. It is suggested that degraded salivary proteins yield di- and oligopeptides, some of which contain arginine. These are rapidly transported into plaque bacteria, in which intracellular peptidases would further cleave the peptides, resulting in production of ammonia and amines and thus stimulating the enzymes in the glycolytic pathway.

4.6. Mineralization of Plaque

In the absence of carbohydrate, the increase in plaque pH (Schroeder, 1969; Kleinberg, 1970) also favors the deposition of calcium and phosphate from saliva into the plaque ecosystem. This mineral deposit, which can occur both supra- and subgingivally, is known as dental calculus or tartar. Dental calculus is thought to contribute to periodontal disease by the irritation of gingival tissue which initiates an inflammatory response.

As early as 1926, Naeslund recognized that filamentous bacteria in dental plaque were involved in the formation of calculus in humans. Subsequently, Wasserman *et al.* (1958) demonstrated that colonies of *A. israelii* were capable of binding calcium independent of metabolic activity. In a microbiological study of calculous dental plaque, Howell *et al.* (1965) found that *A. israelii* and *A. naeslundii* were the predominant organisms in older plaque samples. The same study also showed that streptococci were the predominant organisms in samples of developing calculous plaque, although the direct involvement of these organisms in the process was not demonstrated at that time.

More recently, Sidaway (1978a) characterized the flora of both supra- and subgingival calculus from 20 patients and found it to be more complex than previously suspected. The microflora was qualitatively similar to that of dental plaque in the gingival margin; however, *S. mutans* was absent, while *S. sanguis* appeared in higher numbers. Of 34 species tested for calculus formation in an *in vitro* system, 18 species showed evidence of calcification (Sidaway, 1978b). A significant number of the latter organisms were gram negative, suggesting that they may play a greater role in calculus formation than had been previously indicated.

Zander *et al.* (1960) demonstrated that the densely calcified areas of plaque contained large crystals of hydroxyapatite both between and within micro-organisms. However, intracellular crystals were not observed in less densely calcified areas so that it was not clear whether plaque bacteria were directly involved in the calcification process or passively engulfed by crystal growth (Ennever and Creamer, 1967).

Ennever (1960, 1963), extending the earlier work of Bulleid (1925), demonstrated that *B. matruchotii* was capable of converting calcium salts to an intracellular crystal possessing an X-ray-diffraction pattern of calcium hydroxyapatite. This was confirmed by Takazoe (1961),

Rizzo *et al.* (1963), in an extensive study of calcification by oral bacteria, demonstrated that this process took two forms, one occurring within the cell and another occurring only at the cell surface. The calcification seen with *Actinomyces*, *Bacterionema*, and the streptococci was an intracellular event, while cell-surface calcification was present in *Veillonella* and the diphtheroids. Takazoe *et al.* (1963) further demonstrated that, provided cells of *B. matruchotii* were permitted to grow for a period of more than three days, calcification would occur within nongrowing cells. Subsequent work (Takazoe and Nakamura, 1965) showed that polyphosphates interfered with crystal formation in intact cells of the organism and in cell-free extracts. This has been confirmed recently by Killian and Ennever (1975), who showed that the addition of magnesium to cultures of *B. matruchotii* resulted in the replacement of apatite by an amorphous precipitate.

Strains of *S. mutans* and *S. sanguis* are capable of forming intracellular hydroxyapatite (Streckfuss *et al.*, 1974). Several strains of *S. mutans* were thought to synthesize both intra- and extracellular crystals. However, despite these studies, very little is known about the actual mechanism(s) involved in both the intra- and extracellular calcification processes.

4.7. Inhibitors of Metabolism

The principal natural environmental factor regulating the microbial communities in the oral cavity is saliva. Saliva contains a variety of antibacterial agents, such as lysozyme, lactoferrin, and lactoperoxidase (see review by Mandel, 1976). These are bacteriostatic, and often bactericidal, to oral bacteria under the appropriate conditions. Saliva also contains buffering systems and other components, such as antacids, which serve to control microbial function (Mandel, 1976; Kleinberg *et al.* 1976). However, despite these natural constraining factors, dental caries and periodontal disease are endemic in man.

In order to control these diseases, a variety of antimicrobial agents have been used to reduce the level of microorganisms on either the tooth or gingival surfaces. Since the cariostatic mechanisms of fluoride (Brown and Konig, 1977) and other chemotherapeutic agents (Loesche, 1976a) have been reviewed recently, the discussion here will be restricted to a brief overview of the subject.

Historically, the use of various antimicrobial agents, such as fluoride, is based on the concept that dental caries and periodontal disease result from the activity of the entire microbial flora, since an obvious pathogen(s) has not been isolated. This approach, which Loesche (1976a) refers to as the nonspecific plaque hypothesis (NSPH), implies that the individual microbial populations in the total plaque flora are "pathogenic," and the host has a threshold level for the irritants in plaque. Consequently, the view is held that the simplest method of treatment is to mechanically remove all of the microflora. Since plaque can accumulate to maximum levels within five days (Löe *et al.*, 1965), this is clearly an unsatisfactory approach despite its continued use in dental practice. Contrary to the NSPH is the specific plaque hypothesis (SPH) (Loesche, 1976a), which states that only certain small ecosystems contain "pathogen(s)," which, under the proper environmental conditions, will proliferate to promote the disease. The "pathogen(s)" in this case could, in fact, originate in the indigenous microflora. Therefore, it can be seen that the SPH is analogous to the approach taken with nondental infectious diseases, and the treatment applied is one which reduces or removes the offending pathogen, without necessarily affecting the normal flora.

Various chemical antimicrobial agents are effective antiplaque agents; however, in many cases their delivery to the human population is restricted by ethical, governmental, or mechanical problems. As mentioned by Loesche (1976a), antibiotics such as penicillin, although effective against dental caries (Littleton and White, 1964), cannot be used ethically to control large microbial communities in nonlethal diseases. However, antibiotics such as kanamycin (Loesche *et al.*, 1977) and vancomycin (Kaslick *et al.*, 1973; De Paola *et al.*, 1974) have been used effectively in short-term experiments to control the oral microflora.

The most effective anticaries agent in general use, as confirmed by a variety of epidemiological and clinical studies (Newbrun, 1972; Aasenden *et al.*, 1972; Rugg-Gunn *et al.*, 1973; Backer Dirks, 1974; Brown *et al.*, 1976), is fluoride. A considerable amount of information on the antimicrobial effects of this compound has accumulated in the dental literature. Bibby and van Kesteren (1940) first demonstrated that glycolysis by pure cultures of oral streptococci was inhibited by low concentrations of fluoride. As reviewed by Hamilton (1977), it is clear that fluoride inhibits microbial growth and metabolism in dental plaque (Stralfors, 1950; Jenkins *et al.*, 1969; Woolley and Rickles, 1971; Loesche *et al.*, 1975b; Loesche, 1977; Beighton and MacDougall, 1977) and in saliva (Sandham and Kleinberg, 1969).

The principal mechanism of fluoride action on oral microorganisms is the inhibition of acid production. In the streptococci and probably other glycolytic organisms as well, fluoride inhibits the activity of enolase, which converts 2-phosphoglycerate to phosphoenolpyruvate (PEP) in the glycolytic pathway. This, in turn, results in the inhibition of glucose transport via the PEP phosphotransferase because PEP formation is reduced (Kanapka and Hamilton, 1971;

Hamilton, 1977). Coupled with the reduction in glycolysis in these bacteria, glycogen synthesis is severely restricted due to the unavailability of glucose-6-phosphate (Kanapka *et al.*, 1971).

Fluoride is reported to have a selective effect on certain populations in the plaque ecosystem, but this appears to depend on the method of delivery and the location of the ecosystem. For example, Loesche *et al.* (1975b) have shown that 8-10 daily applications of acidulated phosphate fluoride reduced the concentrations of *S. mutans* in human occlusal plaque but had no effect on the levels of the organism in the approximal plaque. *S. sanguis*, on the other hand, was unaffected in both locations. Similar results were observed in experiments in which rats were ingesting high levels of fluoride (250 μg/ml) in the drinking water (Beighton and MacDougall, 1977). In the latter experiments, the levels of *S. sanguis*, *A. viscosus*, and *Actinobacillus* species either were unaffected or actually increased in the presence of sodium fluoride. Contrary to these findings is evidence that fluoride administered as supplements from birth (van Houte *et al.*, 1978) or in the drinking water (de Stoppelaar *et al*, 1969) does not appreciably affect the colonization of *S. mutans* in dental plaque. Since some oral streptococci readily develop fluoride resistance (Williams, 1967; Hamilton, 1969), it is perhaps unfortunate that these studies did not include a test for this characteristic, since it would have provided valuable information on the adaptability of the plaque ecosystem.

Although it is known that water fluoridation results in a hydroxyapatite less susceptible to acid demineralization (Posner, 1969), considerable speculation exists as to the influence of fluoride on microbial colonization of the tooth surface. Rolla and Melsen (1975) have suggested that fluoride can interfere with the binding of protein, salivary glycoprotein, and bacteria to the tooth surface by replacing acidic protein groups adsorbed to calcium molecules in the hydroxyapatite crystal. Glantz (1969) came to a similar conclusion from a demonstration that fluoride lowered the surface energy of plaque.

There is some indication that fluoride reduces the amount of extracellular polysaccharide (EPS) in plaque receiving sucrose (Broukal and Zajícek, 1974), which could have an influence on the colonization of certain microorganisms. The fluoride effect would appear to be associated with a reduction in overall energy metabolism, since the EPS-synthesizing enzymes, glucosyl and fructosyl transferases, from oral streptococci are unaffected by high concentrations of fluoride (Carlsson *et al*; 1969; Carlsson, 1970; Sharma *et al.*, 1974). Since fluoride increases the ratio of fructose to glucose in EPS following growth on sucrose by *S. mutans* strains (Bowen and Hewitt, 1974), it is conceivable that fluoride is selectively inhibiting the early reactions in the metabolism of fructose by these organisms.

Chlorhexidine is a therapeutic agent which is very effective at low concentrations against the organisms in the plaque ecosystem. Chlorhexidine, as a 0.1% mouthwash used daily, has been shown to reduce plaque weight in both short-

(Davies *et al.*, 1970; Löe and Schiøtt, 1970) and long-term studies (Löe *et al.*, 1976). Salivary bacterial concentrations were reduced 85–90% in short-term (Löe and Schiøtt, 1970) and 30–50% in long-term experiments employing a 0.2% mouthwash (Schiøtt *et al.*, 1976). Significant shifts in the microbial populations were not seen in the latter study, which lasted two years and seven months, although Emilson *et al.* (1976) have reported an alteration in the proportions of *S. mutans* and *S. sanguis* in plaque, with the former organism disappearing on chlorhexidine treatment. Chlorhexidine-resistant mutants in plaque are not seen even after a two-year exposure to the compound (Gjermo and Eriksen, 1974). Chlorhexidine is a positively charged molecule which is readily adsorbed to oral structures and is not rapidly cleared from the oral cavity (Rolla *et al.*, 1970). However, this compound, while being an excellent antimicrobial agent, nevertheless has several drawbacks—it has a bitter taste and leaves a brown stain on the teeth. At this time, it is not approved for general use in North America, but it is available in Europe.

5. Microbial Interactions

Microbial interactions play an important role in the development of the various communities in the oral cavity as they do in other ecological systems. These interactions can be divided into either beneficial (commensalism, protocooperation, or symbiosis) or antagonistic (amensalism) relationships, although many of the interactions are probably neither beneficial nor harmful (neutralism) (Alexander, 1971).

As mentioned at the outset, MacDonald and his co-workers (1954) were probably the first to examine in detail some relationships among oral bacteria. It could be demonstrated in guinea pigs that combinations of pure cultures isolated from the natural flora of periodontal pockets produced necrotic lesions having similar inflammatory and suppurative characteristics to typical fusospirochetal infections. Subsequently (Macdonald *et al.*, 1963), it was shown that, of all the organisms tested, a strain of *Fusobacterium* and an anaerobic diphtheroid were all that was required to support the growth of oral spirochetes in a mixed culture. The associated organisms were shown to reduce the oxidation-reduction potential of the medium as well as provide an unidentified growth factor. Spirochetes would grow without the accompanying growing organisms provided that their culture filtrates and a reducing agent were supplied to reduce the redox potential (Socransky *et al.*, 1964). Cocarboxylase, isobutyrate, and putrescine provided by the associated organisms were shown to be required for maximum growth.

In similar types of experiments, the pathogenicity of *B. melaninogenicus* isolated from periodontal pockets was examined in pure culture, singly and in combination with other oral anaerobes (MacDonald *et al.*, 1960). Originally, it was demonstrated that *Bacteroides* species, particularly *B. melaninogenicus*, were constant members of mixed infections but were usually nonpathogenic in pure culture. The addition of *B. melaninogenicus* to other oral anaerobes that were nonpathogenic either in pure culture or mixed in various combinations produced an infective mixture. The minimum synergistic combination capable of producing a necrotic lesion in a guinea pig contained *B. melaninogenicus*, two other *Bacteroides* strains, and a facultative diphtheroid, which was shown to provide a naphthoquinone essential for the growth of *B. melaninogenicus*. Later (MacDonald and Gibbons, 1962; MacDonald *et al.*, 1963; Socransky and Gibbons, 1965), it was shown that *B. melaninogenicus* was almost always the key pathogen in most anaerobic infections of mucous membranes. A variety of other organisms was shown to be capable of a supporting role by providing factors for the growth and metabolism of *B. melaninogenicus*. Recent evidence indicates that collagenase produced by the latter organism, when combined with *F. fusiforme*, enhances the formation of lesions in rabbits (Kaufman *et al.*, 1972). As pointed out by Loesche (1968), *B. melaninogenicus* is dependent on the host for amino acids and carbohydrates as well as hemin, while the cohabiting organisms provide vitamin K and a reduced environment.

Another group of fastidious oral microbes benefiting from the complex anaerobic flora of plaque is the vibrios. For example, *Vibrio sputorum*, which employs either H_2 or formate for energy *in vitro* (Loesche, 1968), could obtain H_2 *in vivo* from the metabolism of lactate by *Veillonella* species (Rogosa, 1964). Formate could be provided by the streptococci (Carlsson and Griffith, 1974) or by *B. oralis*, *Fusobacterium nucleatum*, or *B. melaninogenicus* (Loesche and Gibbons, 1968).

As mentioned previously, extracellular polysaccharides do benefit various microorganisms in the oral ecosystem by providing an energy source upon degradation. In addition, the lactate produced by acidogenic microbes is utilized as an energy source by *Veillonella* species (Mikx and van der Hoeven, 1975). Another example of a synergistic relationship is evident in the demonstration by Carlsson (1971) that *S. sanguis*, when grown in a mixed culture with *S. mutans*, could provide the latter organism with *p*-aminobenzoic acid, which was essential for growth.

Apart from nutritional interrelationships, it is also apparent that various organisms in an oral ecosystem that colonize the tooth efficiently can assist in the colonization by other microbial types having less affinity for this surface. For example, an adherent strain of *S. salivarius* was shown to mediate the adherence of *Veillonella* and several streptococcal species that do not normally colonize smooth surfaces (McCabe and Donkersloot, 1977). Also, the appearance of "corn-cob" formations in plaque (Listgarten *et al.*, 1973) suggests the association of filamentous organisms and small coccoid forms in a similar process. Takazoe *et al.* (1978) were able to form "corn cobs" *in vitro* by combining *B. matruchotii* with various strains of *S. sanguis*. The colonizing potential of combinations of oral microorganisms has been studied *in vitro* with Nichrome wires suspended in liquid medium. For example, "plaque" formation by mixtures containing *S. mutans* or *S. sanguis* and *L. casei* was reduced significantly compared to that formed either by streptococcal strain alone (Miller and Kleinman, 1974). The combination of *S. mutans* and *Candida albicans*, on the other hand, resulted in higher surface aggregates those produced by *S. mutans* alone. The results from this test system must be interpreted with caution, however, since the colonization of mouse epithelial cells by *C. albicans* is suppressed by *S. salivarius* and *S. mitis* and by the indigenous flora in human saliva (Liljemark and Gibbons, 1973). A strain of *S. salivarius* was shown to be inhibitory to a variety of streptococcal species, anaerobic gram-positive cocci, and *Corynebacterium diphtheriae*, *C. hofmanii*, and *C. xerosis* (Bill and Washington, 1975).

The formation of inhibitory substances by members of the microbial community is an important factor in the ecology of the oral cavity. Weerkamp *et al.* (1977) examined the production of antagonistic substances by 69 strains of *Streptococcus* freshly isolated from human dental plaque. *S. mutans* strains produced a bacteriocin-like substance which was relatively inactive against ecologically unrelated organisms such as *E. coli*, *Staphylococcus aureus*, and *S. faecalis*, as well as other strains of *S. mutans*, but was active against plaque organisms such as *S. sanguis*, *A. viscosus*, and *B. matruchotii*. *S. sanguis*, in this study, was shown to be a relatively inactive antagonist. Nutrition had a variable effect on the inhibitory factors depending on the strains employed. A variety of antagonistic substances was produced by the test organisms, most of which were sensitive to proteolytic enzymes, and these substances were classified as bacteriocins. A small number were nonproteins and others were lipid containing, since they were sensitive to lipolytic enzymes. Hydrogen peroxide, known to be produced by *S. sanguis* and active against a variety of oral microorganisms under aerobic conditions (Holmberg and Hallander, 1973), was not a factor, since the experiments were performed anaerobically. However, the inhibitory effects of metabolic acids could not be totally discounted, since it is known that oral bacteria, particularly the streptococci, produce inhibitory levels of acid end products when exposed to carbohydrate diets (Donoghue and Tyler, 1975a, b).

The ability of *S. sanguis* to predominate during early plaque formation has been attributed to its production of bactericidal concentrations of hydrogen peroxide, since the environment is largely aerobic. With time, the action of salivary catalase and peroxidase would eventually permit colonization of other organisms, producing an anaerobic environment and thereby removing the selective advantage of *S. sanguis* (Holmberg and Hallander, 1973).

Bacteriocin production by oral microorganisms is now well established.

Streptococci (Kelstrup and Gibbons, 1969; Kelstrup *et al.* 1971), particularly strains of *S. mutans* (Berkowitz and Jordan, 1975; Rogers, 1975, 1976a) and *S. sanguis* (Nakamura *et al.*, 1977), are thought to be a major source of bacteriocins in dental plaque. Rogers (1976b) has reported that 70% of the 143 strains of *S. mutans* he tested were bacteriocinogenic against a variety of indicator strains including other streptococci, enterococci, and a number of unrelated gram-positive bacteria. Many strains appeared to produce more than one type of bacteriocin with considerable diversity among the various types. Both *in vitro* (Weerkamp *et al.*, 1977) and *in vivo* (van der Hoeven and Rogers, 1978; Rogers *et al.*, 1978) mixed-culture experiments have demonstrated that bacteriocin-producing strains have a selective advantage when in competition with non-bacteriocin-producing organisms.

One of the more interesting approaches to the study of the interactions of oral bacteria has been that of Mikx and van der Hoeven in Nijmegen, The Netherlands. These workers have used gnotobiotic and specific-pathogen-free (SPF) rats in model systems to examine plaque development and metabolism *in vivo*. They have related the *in vivo* findings to controlled *in vitro* systems using continuous culture (Mikx and van der Hoeven, 1975; Mikx *et al.*, 1976a; van der Hoeven, 1976). In early experiments, it was shown that diassociation of bacteria in the mouths of rats could modify caries attack (Mikx *et al.*, 1972), and this research was extended further to a wider range of oral bacteria (Mikx *et al.*, 1975a, 1976b). Similarly, the sequence of inoculation of bacterial strains into SPF rats influenced their ability to establish (Mikx *et al.*, 1975b).

The use of this *in vivo* model system linked to continuous-culture studies must be regarded as one of the most important in the laboratory study of oral ecosystems. The early results of the experiments provided some evidence for the proposals of other workers on mechanisms of oral ecology, e.g., the uptake of lactate by oral microorganisms reducing caries activity (Douglas, 1950). Recently, isotachophoresis has been introduced as a sensitive method to measure organic acids in small samples of plaque (Dellebarre *et al.*, 1977). The application of this technique to the *in vivo* model and to continuous-culture studies may allow the workers at Nijmegen to be the first to determine precisely the relationship between the composition of the plaque community and its acid products.

Another experimental approach to the study of microbial interactions has been the use of artificial fissures mounted onto teeth in human subjects (Svanberg and Loesche, 1977; Mikx and Svanberg, 1978). The fissures were placed either in a sterile condition or precolonized *in vitro* with a labeled organism. The sterile artificial fissures employed by Svanberg and Loesche (1977) were colonized with 10^7 colony-forming units (CFUs) in 1 day, and this value was shown to be relatively constant for periods up to 21 days. Despite this constancy in the total numbers, the concentrations of *S. mutans* and *S. sanguis* increased or decreased with time depending on the individuals. Initial colonization was shown to depend on the concentration of the respective organism in saliva. For example, if the concentration of *S. mutans* in saliva was 10^3 CFUs, the organism would colonize the fissure and increase in number to 6% of the total community. However, the insertion of the fissures during regimens reducing the salivary concentrations of *S. mutans* could result in failure of the organism to colonize the fissure. The organism would continue to be excluded even during postregimen periods when *S. mutans* was allowed to increase to levels associated with colonization. These studies, therefore, indicate the importance of the pioneer microflora to the composition of the fissure ecosystem.

6. Summary

This review has had to be limited in its scope, and for this reason several topics have been omitted or abbreviated. These include experiments with artificial mouth systems (Dibdin *et al.*, 1976), animal models (Gordon and Pesti, 1971; van der Hoeven *et al.*, 1972) and studies of the role of genetic factors in oral communities. The latter topic has received some attention recently in the search for plasmids coding for some activities in oral streptococci (Clewell *et al.*, 1976) and lactobacilli (Hammond and Darkes, 1976) and for mutants to aid in understanding the mechanisms of pathogenesis (Johnson *et al.*, 1974; Raina and Ravin, 1976; Freedman *et al.*, 1976). Although immune mechanisms have been discussed, the subject of vaccination against caries has not been pursued in detail. However, sufficient references to this topic have been included to enable those interested to explore further. Little has been said of the mechanisms of enamel destruction or of the breakdown of gingival tissue. These subjects are complex and are perhaps better placed in reviews of pathogenic mechanisms rather than microbial ecology.

The consideration of the mouth as an ecosystem and of the application of ecological principles to oral problems is a relatively recent phenomenon among oral microbiologists. It might be fitting to end this chapter by giving credit to those workers at the Forsyth Dental Center, Boston, Massachusetts, and the University of Nijmegen, The Netherlands, who have pioneered work in this area.

References

Aasenden, R., De Paola, P. F., and Brudevold, F., 1972, Effects of daily rinsing and ingestion of fluoride solutions upon dental caries and enamel fluoride, *Arch. Oral Biol.* **17:** 1705–1714.

Alexander, M., 1971, *Microbial Ecology*, John Wiley and Sons, New York.

Alshamony, L., Goodfellow, M., Minnikin, D. E., Bowden, G. H., and Hardie, J. M., 1977, Fatty and mycolic acid composition of *Bacterionema matruchotti* and related organisms, *J. Gen. Microbiol.* 98:205-213.

Backer Dirks, O., 1974, The benefits of water fluoridation, *Caries Res. Suppl.* 8:2-15.

Baier, R. E., 1977, On the formation of biological films, *Swed. Dent. J.* 1:261-271.

Baker, J. J., Chan, S. P., Socransky, S. S., Oppenheim, J. J., and Mergenhagen, S. E., 1976, Importance of *Actinomyces* and certain gram-negative anaerobic organisms in the transformation of lymphocytes from patients with periodontal disease, *Infect. Immun.* 13:1363-1368.

Beighton, D., and Colman, G., 1976, A medium for the isolation of Actinomycetaceae from human dental plaque, *J. Dent. Res.* 55:875-878.

Beighton, D., and MacDougall, W. A., 1977, The effects of fluoride on the percentage bacterial composition of dental plaque, on caries incidence, and on the *in vitro* growth of *Streptococcus mutans, Actinomyces viscosus,* and *Actinobacillus* sp., *J. Dent. Res.* 56:1185-1191.

Beighton, D., and Miller, W. A., 1977, A microbiological study of normal flora of macropod dental plaque, *J. Dent. Res.* 56:995-1000.

Berg, R. D., and Savage, D. C., 1975, Immune responses of specific pathogen free and gnotobiotic mice to antigens of indigenous and nonindigenous microorganisms, *Infect. Immun.* 11:320-329.

Berger, U., 1967, Zur Systematik der Neisseriaceae, *Zentralbl. Bakteriol. Parasitenkd. Infektionskr. Hyg. Abt. 1: Orig.* 205:241-248.

Berger, U., and Catlin, B. W., 1975, Biochemische Unterscheidung von *Neisseria sicca* und *Neisseria perflava, Zentralbl. Bakteriol. Parasitenkd. Infektionskr. Hyg. Abt. 1: Orig. Reihe A* 232:129-130.

Berger, U., and Wulf, B., 1961, Untersuchungen an saprophytischen Neisserien, *Zentralbl. Bakteriol. Parasitenkd. Infektionskr. Hyg. Abt. 1* 147:257-268.

Berglund, S. E., Rizzo, A. A., and Mergenhagen, S. E., 1969, The immune response in rabbits to bacterial somatic antigen administered via the oral mucosa, *Arch. Oral Biol.* 14:7-17.

Berkowitz, R. J., and Jordan, H. V., 1975, Similarity of bacteriocins of *Streptococcus mutans* from mother and infant, *Arch. Oral Biol.* 20:725-730.

Berman, K. S., and Gibbons, R. J., 1966, Iodophilic polysaccharide synthesis by human and rodent oral bacteria, *Arch. Oral Biol.* 11:533-542.

Berman, K. S., Gibbons, R. J., and Nalbandian, J., 1967, Localization of intracellular polysaccharide granules in *Streptococcus mitis, Arch. Oral Biol.* 12:1133-1138.

Beveridge, T. J., and Goldner, M., 1973, Statistical relationships between the presence of human subgingival anaerobic diphtheroids and periodontal disease, *J. Dent. Res.* 52:451-453.

Bibby, B. G., 1976, Influence of diet on the bacterial composition of plaques, in: *Microbial Aspects of Dental Caries* (H. M. Stiles, W. J. Loesche, and T. C. O'Brien, eds.), *Microbiol. Abstr. Spec. Suppl.* 2:477-490.

Bibby, B. G., and van Kesteren, M., 1940, The effect of fluorine on mouth bacteria, *J. Dent. Res.* 19:391-401.

Bill, N. J., and Washington, J. A., 1975, Bacterial interference by *Streptococcus salivarius, Am. J. Clin. Pathol.* 64:116-120.

Black, G. V., 1899, Susceptibility and immunity in dental caries, *Dent. Cosmos* 41:826-830.

Blank, C. H., and Georg, L. K., 1968, The use of fluorescent antibody methods for the detection and identification of *Actinomyces* species in clinical material, *J. Lab. Clin. Med.* 71:283-293.

Bovre, K., 1967, Transformation and DNA base composition in taxonomy, with special reference to recent studies in *Moraxella* and *Neisseria*, *Acta Pathol. Microbiol. Scand.* 69:123–144.

Bowden, G. H., 1969, The components of the cell walls and extracellular slime of four strains of *Staphylococcus salivarius* isolated from human dental plaque, *Arch. Oral Biol.* 14:685–697.

Bowden, G. H., and Hardie, J. M., 1971, Anaerobic organisms from the human mouth, in: *Isolation of Anaerobes* (D. A. Shapton and R. G. Board, eds.), pp. 177–205, Academic Press, London.

Bowden, G. H., and Hardie, J., 1973, Commensal and pathogenic *Actinomyces* species in man, in: *Actinomycetales: Characteristics and Practical Importance* (G. Sykes and F. A. Skinner, eds.), pp. 277–299, Academic Press, London.

Bowden, G. H., and Hardie, J. M., 1978, Gram-positive pleomorphic (coryneform) organisms from the mouth, in: *Coryneform Bacteria* (I. J. Bousefield and A. G. Callely, eds.), pp. 235–263, Academic Press, London.

Bowden, G. H., Hardie, J. M., and Slack, G. L., 1975, Microbial variations in approximal dental plaque, *Caries Res.* 9:253–277.

Bowden, G. H., Hardie, J. M., McKee, A. S., Marsh, P. D., Fillery, E. D., and Slack, G. L., 1976, The microflora associated with developing carious lesions of the distal surfaces on the upper first premolars in 13–14 year old children, in: *Microbial Aspects of Dental Caries* (H. J. Stiles, W. J. Loesche, and T. C. O'Brien, eds.), *Microbiol. Abstr. Spec. Suppl.* 1:223–241.

Bowden, G. H., Hardie, J. M., Fillery, E. D., Marsh, P. D. and Slack, G. L., 1978, Microbial analyses related to caries susceptibility, in: *Methods of Caries Prediction* (Workshop Proceedings) (B. G. Bibby and R. J. Shern, eds.), *Microbiol. Abstr. Spec. Suppl.* pp. 83–97, Information Retrieval, Inc., Washington, D.C.

Bowen, W. H., 1969, A vaccine against dental caries. A pilot experiment with monkeys (*Macaca irus*), *Br. Dent. J.* 126:159–160.

Bowen, W. H., 1976, Nature of plaque, *Oral Sci. Rev.* 9:3–22.

Bowen, W. H., and Hewitt, M. J., 1974, Effect of fluoride on extracellular polysaccharide production by *Streptococcus mutans*, *J. Dent. Res.* 53:627–629.

Bowen, W. H., Cohen, B., and Colman, G., 1975, Immunisation against dental caries, *Br. Dent. J.* 139:45–58.

Bowen, W. H., Genco, R. J., and O'Brien, T. (eds.), 1976, *Immunologic Aspects of Dental Caries*, *Immunol. Abstr. Spec. Suppl.*, Information Retrieval, Inc., Washington, D.C.

Bramstedt, F., and Lusty, C. J., 1968, The nature of the intracellular polysaccharides synthesized by streptococci in the dental plaque, *Caries Res.* 2:201–213.

Brandtzaeg, P., 1976, Synthesis and secretion of secretory immunoglobulins: With special references to dental diseases, *J. Dent. Res. Spec. Iss. C* 55:102–114.

Brandtzaeg, P., and Tolo, K., 1977, Immunoglobulin systems of the gingiva, in: *The Borderland between Caries and Periodontal Disease* (T. Lehner, ed.), pp. 145–183, Academic Press, London.

Brandtzaeg, P., Fjellanger, I., and Gjeruldsen, S. T., 1968, Adsorption of Immunoglobulin A onto oral bacteria "*in vivo*," *J. Bacteriol.* 96:242–249.

Bratthall, D., and Gibbons, R. J., 1975, Antigenic variation of *S. mutans* colonizing gnotobiotic rats, *Infect. Immun.* 11:1231–1236.

Brooks, J. B., Kellogg, D. S., Thacker, L., and Turner, E. M., 1971, Analysis by gas chromatography of fatty acids found in whole cultural extracts of *Neisseria* species, *Can. J. Microbiol.* 17:531–543.

Broukal, Z., and Zajicek, O., 1974, Amount and distribution of extracellular polysaccharides in dental microbial plaque, *Caries Res.* 8:97–104.

Brown, A. T., and Wittenberger, C. L., 1972, Fructose-1,6-diphosphate-dependent lactate dehydrogenase from a cariogenic *Streptococcus*: Purification and regulatory properties, *J. Bacteriol.* 110:604–615.

Brown, L. R., Handler, S., Allen, S. S., Shea, C., Wheatcroft, M. G., and Frome, W. J., 1973, Oral microbial flora of the marmoset, *J. Dent. Res.* 52:815–822.

Brown, L. R., Dreizen, S., and Handler, S., 1976, Effects of selected caries preventive regimens on microbial changes following irradiation-induced zerostomia in cancer patients, in: *Microbial Aspects of Dental Caries* (H. M. Stiles, W. J. Loesche, and T. C. O'Brien, eds.), *Microbial Abstr. Spec. Suppl.* 1:275–290.

Brown, W. E., and Konig, K. G., 1977, Cariostatic mechanisms of fluorides, *Caries Res.* 11: 1–327.

Brown, W. R., and Lee, E., 1974, Radioimmunological measurements of bacterial antibodies, *Gastroenterology* 66:1145–1153.

Buchanan, B. B., and Pine, L., 1963, Factors influencing the fermentation and growth of an atypical strain of *Actinomyces naeslundii*, *Sabouraudia* 3:26–39.

Buchanan, B. B., and Pine, L., 1967, Path of glucose breakdown and cell yields of a facultative anaerobe *Actinomyces naeslundii*, *J. Gen. Microbiol.* 46:225–236.

Buchanan, R. E., and Gibbons, N. E., (eds.), 1974, *Bergey's Manual of Determinative Bacteriology*, Williams and Wilkins, Baltimore.

Bulleid, A., 1925, An experimental study of *Leptothrix buccalis*, *Br. Dent. J.* 46:289–300.

Bunting, R. W., Nicherson, G., Hard, D., and Crowley, M., 1928, Further studies of the relation of *Bacillus acidophilus* to dental caries, II, *Dent. Cosmos* 70:1–15.

Burckhardt, J. J., 1978, Rat memory T lymphocytes: *In vitro* proliferation induced by antigens of *Actinomyces viscosus*, *Scand. J. Immunol.* 7:167–172.

Caldwell, J., Challacombe, S. J., and Lehner, T., 1977, A sequential bacteriological and serological investigation of rhesus monkeys immunized against dental caries with *Streptococcus mutans*, *J. Med. Microbiol.* 10:213–224.

Carlsson, J., 1967, Presence of various types of non-hemolytic streptococci in dental plaque and in other sites of the oral cavity of man, *Odontol. Revy* 18:55–74.

Carlsson, J., 1970, A levansucrase from *Streptococcus mutans*, *Caries Res.* 4:97–113.

Carlsson, J., 1971, Growth of *Streptococcus mutans* and *Streptococcus sanguis* in mixed culture, *Arch. Oral Biol.* 16:963–965.

Carlsson, J., and Gothefors, L., 1975, Transmission of *Lactobacillus jensenii* and *Lactobacillus acidophilus* from mother to child at delivery, *J. Clin. Microbiol.* 1:124–128.

Carlsson, J., and Griffith, C. J., 1974, Fermentation products and bacterial yields in glucose-limited and nitrogen-limited cultures of streptococci, *Arch. Oral Biol.* 19:1105–1109.

Carlsson, J., and Johansson, J., 1973, Sugar and the production of bacteria in the human mouth, *Caries Res.* 7:273–282.

Carlsson, J., Newbrun, E., and Krasse, B., 1969, Purification and properties of dextransucrase from *Streptococcus sanguis*, *Arch. Oral Biol.* 14:469–478.

Carlsson, J., Graham, H., and Jonsson, G., 1975, Lactobacilli and streptococci in the mouth of children, *Caries Res.* 9:333–339.

Challacombe, S. J., and Lehner, T., 1976, Serum and salivary antibodies to cariogenic bacteria in man, *J. Dent. Res.* 55:139–148.

Charlton, G., Fitzgerald, D. B., and Keyes, P. H., 1971, Hydrogen ion activity in dental plaques of hamsters during metabolism of sucrose, glucose and fructose, *Arch. Oral Biol.* 16:655–661.

Chet, I., and Mitchell, R., 1976, Ecological aspects of microbial chemotactic behavior, *Annu. Rev. Microbiol.* 30:221-239.

Clark, W. B., and Gibbons, R. J., 1977, Influence of salivary components and extracellular polysaccharide synthesis from sucrose on the attachment of *Streptococcus mutans* 6715 to hydroxyapatite surfaces, *Infect. Immun.* 18:514-523.

Clark, W. B., Bamman, L. L., and Gibbons, R. J., 1978, Comparative estimates of bacterial affinities and adsorption sites on hydroxyapatite surfaces, *Infect. Immun.* 19:846-853.

Clarke, J. K., 1924, On the bacterial factor in the aetiology of dental caries, *Br. J. Exp. Pathol.* 5:141-146.

Clewell, D. B., Oliver, D. R., Dunny, G. M., Franke, A. E., Yagi, Y., van Houte, J., and Brown, B. L., 1976, Plasmids in cariogenic streptococci, in: *Microbial Aspects of Dental Caries* (H. M. Stiles, W. J. Loesche, and T. C. O'Brien, eds.), *Microbiol. Abstr. Spec. Suppl.* 3:713-724.

Cole, M. F., Arnold, R. R., Mestecky, J., Prince, S., Kulhavy, R., and McGhee, J. R., 1976, Studies with human lactoferrin and *Streptococcus mutans*, in: *Microbial Aspects of Dental Caries* (H. M. Stiles, W. J. Loesche, and T. C. O'Brien, eds.), *Microbiol. Abstr. Spec. Suppl.* 2:359-373.

Colman, G., and Williams, R. E. O., 1972, Taxonomy of some human viridans streptococci, in: *Streptococci and Streptococcal Diseases* (L. W. Wanamaker, and J. M. Matsu, eds.), pp. 281-299, Academic Press, New York.

Cornick, D. E. R., and Bowen, W. H., 1971, Development of the oral flora in newborn monkeys (*Macaca irus*), *Br. Dent. J.* 130:231-234.

Cowman, R. A., Perrella, M. M., Adams, B. O., and Fitzgerald, R. J., 1975, Amino acid requirements and proteolytic activity of *Streptococcus sanguis*, *Appl. Microbiol.* 30:374-480.

Coykendall, A. L., 1976, On the evolution of *Streptococcus mutans* and dental caries, in: *Microbial Aspects of Dental Caries* (H. M. Stiles, W. J. Loesche, and T. C. O'Brien, eds.), *Microbiol. Abstr. Spec. Suppl.* 3:703-712.

Coykendall, A. L., Specht, P. A., and Samol, H. H., 1974, *Streptococcus mutans* in a wild, sucrose-eating rat population, *Infect. Immun.* 10:216-219.

Crawford, A., Socransky, S. S., and Bratthall, G., 1975, Predominant cultivable microbiota of advanced periodontitis, *J. Dent. Res. Spec. Iss. A* 54:Abstract No. 209.

Critchley, P., 1969, The breakdown of the carbohydrate and protein matrix of dental plaque, *Caries Res.* 3:249-265.

Critchley, P., Wood, J. M., Saxton, C. A., and Leach, S. A., 1967, The polymerisation of dietary sugars by dental plaque, *Caries Res.* 1:112-129.

Cummins, C. S., 1975, Identification of *Propionibacterium acnes* and related organisms by precipitin tests with trichloracetic acid extracts, *J. Clin. Microbiol.* 2:104-110.

Cummins, C., and Johnson, J. L., 1974, *Corynebacterium parvum*: A synonym for *Propionibacterium acnes*, *J. Gen. Microbiol.* 80:433-442.

Da Costa, T., and Gibbons, J. R., 1968, Hydrolysis of levan by mouth plaque streptococci, *Arch. Oral Biol.* 13:609-618.

Da Costa, T., Bier, L. C., and Gaida, F., 1974, Dextran hydrolysis by a *Fusobacterium* strain isolated from human dental plaque, *Arch. Oral. Biol.* 19:341-342.

Darwish, S., Hyppa, T., and Socransky, S. S., 1978, Studies of the predominant cultivable microbiota of early periodontitis, *J. Periodontal Res.* 13:1-16.

Dawes, E. A., and Ribbons, D. W., 1964, Some aspects of the endogenous metabolism of bacteria, *Bacteriol. Rev.* 28:1126-1149.

Davies, R. M., Borglum-Jensen, S. C., Rindom Schiøtt, C., and Löe, H., 1970, The effect of topical application of chlorhexidine on the bacterial colonization of teeth and gingiva, *J. Periodontal Res.* 5:96–101.

Dellebarre, C. W., Franken, H. C. M., Camp, P. J. M., and van der Hoeven, J. S., 1977, Determination of organic acid in dental plaque by isotachophoresis, *J. Dent. Res. Spec. Iss. A* 56:Abstract No. 313.

Dent, V. E., Hardie, J. M., and Bowden, G. H., 1976, A preliminary study of dental plaque on animal teeth, *J. Dent. Res. Spec. Iss. D* 55:Abstract No. 85D.

Dent, V. E., Hardie, J. M., and Bowden, G. H., 1978, Streptococci isolated from dental plaque of animals, *J. Appl. Bacteriol.* 44:249–258.

De Paola, P. F., Jordan, H. V., and Berg, J., 1974, Temporary suppression of *Streptococcus mutans* in humans through topical application of vancomycin, *J. Dent. Res.* 53:108–114.

de Stoppelaar, J. D., van Houte, J., and Backer Dirks, O., 1969, The relationship between extracellular polysaccharide-producing streptococci and smooth surface caries in 13-year old children, *Caries Res.* 3:190–199.

de Stoppelaar, J. D., van Houte, J., and Backer Dirks, O., 1970, The effect of carbohydrate restriction on the presence of *Streptococcus mutans, Streptococcus sanguis* and iodophilic polysaccharide-producing bacteria in human dental plaque, *Caries Res.* 4:114–123.

Dewar, M. G., and Walker, G. J., 1975, Metabolism of the polysaccharides of human dental plaque, *Caries Res.* 9:21–35.

Dibdin, G. H., Shellis, R. P., and Wilson, C. M., 1976, An apparatus for the continuous culture of microorganisms on solid surfaces with special reference to dental plaque, *J. Appl. Bacteriol.* 40:261–268.

Dipersio, T. R., Mattingly, S. J., Higgins, M. L., and Shockman, G. D., 1974, Measurement of intracellular polysaccharide in two cariogenic strains of *Streptococcus mutans* by cytochemical and chemical methods, *Infect. Immun.* 19:597–604.

Dirksen, T. R., Little, M. F., and Bibby, B. G., 1963, The pH of carious cavities II. The pH at different depths in isolated cavities, *Arch. Oral Biol.* 8:91–97.

Dobbs, E. C., 1932, Local factors in dental caries, *J. Dent. Res.* 12:853–856.

Donoghue, H. D., and Tyler, J. E., 1975a, Antagonisms amongst streptococci isolated from the human oral cavity, *Arch. Oral Biol.* 20:381–387.

Donoghue, H. D., and Tyler, J. E., 1975b, Role of lactic and acetic acids in microbial antagonism, *Caries Res.* 9:322.

Douglas, H. C., 1950, On the occurrence of the lactate fermenting anaerobe *Micrococcus lactilyticus* in human saliva, *J. Dent. Res.* 29:304–306.

Dreizen, S., and Brown, L. R., 1976, Xerostomia and dental caries, in: *Microbial Aspects of Dental Caries* (H. M. Stiles, W. J. Loesche, and T. C. O'Brien, eds.), *Microbiol. Abstr. Spec. Suppl.* 1:263–273.

Drucker, D. B., and Melville, T. H., 1968, Fermentation end-products of cariogenic and non-cariogenic streptococci, *Arch. Oral Biol.* 13:563–570.

Duchin, S., and van Houte, J., 1978, Colonization of teeth in humans by *Streptococcus mutans* as related to its concentration in saliva and host age, *Infect. Immun.* 20:120–125.

Dvarskas, R. A., and Coykendall, A. L., 1975, *Streptococcus mutans* in wild rats on a low sucrose diet, *J. Dent. Res. Spec. Iss. A* 54:Abstract No. 330.

Edgar, W. M., 1976, The role of saliva in the control of pH changes in human dental plaque, *Caries Res.* 10:241–254.

Edwardsson, S., 1968, Characteristics of caries-inducing human streptococci resembling *Streptococcus mutans, Arch. Oral Biol.* 13:637–646.

Edwardsson, S., 1974, Bacteriological studies on deep areas of carious dentine, *Odontol. Revy* 25(Suppl. 32):1–143.

Ellen, R. P., 1976, Microbiological assays for dental caries and periodontal disease susceptibility, *Oral Sci. Rev.* 8:3–23.

Ellen, R. P., and Onose, H., 1978, pH Measurements of *Actinomyces viscosus* colonies grown on media containing dietary carbohydrates, *Arch. Oral Biol.* 23:105–109.

Ellwood, D. C., 1976, Chemostat studies of oral bacteria, in: *Microbial Aspects of Dental Caries* (H. M. Stiles, W. J. Loesche, and T. C. O'Brien, eds.), *Microbiol. Abstr. Spec. Suppl.* 3:785–798.

Ellwood, D. C., Hunter, J. R., and Longyear, V. M. C., 1974, Growth of *Streptococcus mutans* in a chemostat, *Arch. Oral Biol.* 19:659–665.

Ellwood, D. C., Hunter, J. C. and Longyear, V. M. C., 1977, Growth of *Actinomyces viscosus* in a chemostat, *J. Dent. Res.* 56:311.

Ellwood, D. C., Phipps, P. J., and Hamilton, I. R., 1979, Effect of growth rate and glucose concentration on the activity of the phospho-enolpyruvate phosphotransferase system in *Streptococcus mutans* grown in continuous culture, *Infect. Immun.* 23:224–231.

Emilson, C. S., Krasse, B., and Westergren, G., 1976, Effect of a fluoride-containing chlorhexidine gel on bacteria in human plaque, *Scand. J. Dent. Res.* 84:56–62.

Emmings, F. G., Evans, R. T., and Gemco, R. T., 1975, Antibody response in the parotid fluid and serum of irus monkeys (*Macaca fascicularis*) after local immunization with *Streptococcus mutans, Infect. Immun.* 12:281–292.

Ennever, J., 1960, Intracellular calcification of oral filamentous organisms, *J. Periodontol.* 31:304–307.

Ennever, J., 1963, Microbiologic calcification, *Ann. N. Y. Acad. Sci.* 109:4–13.

Ennever, J., and Creamer, H., 1967, Microbiological calcification: Bone mineral and bacteria, *Calcif. Tissue Res.* 1:87–93.

Enright, J. J., Friesell, H. E., and Trescher, M. O., 1932, Studies of the cause and nature of dental caries, *J. Dent. Res.* 12:759–827.

Ericson, T., Carlen, A., and Dagerskag, E., 1976, Salivary aggregating factors, in: *Microbial Aspects of Dental Caries* (H. M. Stiles, W. J. Loesche, and T. C. O'Brien, eds.), *Microbiol. Abstr. Spec. Suppl.* 1:151–162.

Evans, R. T., Spaeth, S., and Mergenhagen, S. E., 1966, Bacteriocidal antibody in mammalian serum to obligatory aerobic Gram-negative bacteria, *J. Immunol.* 97:112–119.

Evans, R. T., Emmings, F. G. and Gemco, R. T., 1975, Prevention of *Streptococcus mutans* infection of tooth surfaces by salivary antibody in Irus monkeys (*Macaca fascicularis*), *Infect. Immun.* 12:293–302.

Fahr, A. M., and Berger, U., 1975, Wie anspruchslos sind die sog. anspruchslosen *Neisserien, Zentralbl. Bakteriol. Parasitenkd. Infektionskr. Hyg. Abt. 1: Orig. Reihe A* 230: 551–555.

Fejerskov, O., Theilade, E., Karring, T., and Theilade, J., 1977, Plaque and caries development in experimental human fissures. Structural and microbiological features, *J. Dent. Res. Spec. Iss. A* 56:Abstract No. 456.

Fillery, E. D., Bowden, G. H., and Hardie, J. M., 1978, A comparison of strains of bacteria designated *Actinomyces viscosus* and *Actinomyces naeslundii, Caries Res.* 12:299–312.

Finegold, S. M., and Barnes, E. M., 1977, Report of the ICSB taxonomic subcommittee on gram-negative anaerobic rods, *Int. J. Syst. Bacteriol.* 27:388–391.

Foo, M. C., and Lee, A., 1974, Antigenic cross reaction between mouse intestine and a member of the autochthonous microflora, *Infect. Immun.* 9:1066–1069.

Foo, M. C., Lee, A., and Cooper, G. N., 1974, Natural antibodies and the intestinal flora of rodents, *Aust. J. Exp. Biol. Med. Sci.* 52:321–330.

Foubert, E. L., Jr., and Douglas, H. C., 1948, Studies on the anaerobic micrococci. II. The fermentation of lactate by *Micrococcus lactilyticus, J. Bacteriol.* 56:35–36.

Fox, R. H., and McClain, D. E., 1974, Evaluation of the taxonomic relationship of *Micrococcus cryophilus, Branhomella catarrhalis*, and *Neisseria* by comparative polyacrylamide gel electrophoresis of soluble protein, *Int. J. Syst. Bacteriol.* 24:172–176.

Freedman, M. L., and Coykendall, A. L., 1975, Variation in internal polysaccharide synthesis among *Streptococcus mutans* strains, *Infect. Immun.* 12:475–479.

Freedman, M. L., and Tanzer, J. M., 1974, Dissociation of plaque formation from glucan-induced agglutination in mutants of *Streptococcus mutans, Infect. Immun.* 10:189–196.

Freedman, M. L., Tanzer, J. M., and Eifert, R. L., 1976, Isolation and characterization of mutants of *Streptococcus mutans* with defects related to intracellular polysaccharide, in: *Microbial Aspects of Dental Caries* (H. M. Stiles, W. J. Loesche, and T. C. O'Brien, eds.), *Microbiol. Abstr. Spec. Suppl.* 3:581–596.

Friberg, S., 1977, Colloidal phenomena encountered in the bacterial adhesion to the tooth surface, *Swed. Dent. J.* 1:207–214.

Frostell, G., 1973, Effects of mouth rinses with sucrose, glucose, fructose, lactose, sorbitol and lycasin, *Odontol. Revy* 24:217–223.

Fukui, Y., Fukui, K., and Moryama, T., 1973, Source of neuraminidase in human whole saliva, *Infect. Immun.* 8:329–334.

Gawronski, T. H., Statt, R. A., Zaki, H. A., Harris, R. S., and Folke, L. E. A., 1975, Effects of dietary sucrose levels on extracellular polysaccharide metabolism of human dental plaque, *J. Dent. Res.* 54:881–890.

Geddes, D. A. M., 1972, The production of L(+) and D(–) lactic acid and volatile acids by human dental plaque and the effect of plaque buffering and acidic strength on pH, *Arch. Oral Biol.* 17:537–545.

Geddes, D. A. M., 1975, Acids produced by human dental plaque metabolism *in situ, Caries Res.* 9:98–109.

Geddes, D. A. M., and Jenkins, G. N., 1974, Intrinsic and extrinsic factors influencing the flora of the mouth, in: *The Normal Microbial Flora of Man* (F. A. Skinner and J. G. Carr, eds.), pp. 85–100, Academic Press, London.

Genco, R. J. (ed.), *Immunological Aspects of Dental Caries (Symposium), J. Dent. Res. Spec. Iss. C* 55:c1–c230.

Gibbons, R. J., 1964a, Bacteriology of caries, *J. Dent. Res.* 43:1021–1028.

Gibbons, R. J., 1964b, Metabolism of intracellular polysaccharide by *Streptococcus mitis* and its relation to inducible enzyme formation, *J. Bacteriol.* 87:1512–1520.

Gibbons, R. J., and Banghart, S. B., 1967, Synthesis of extracellular dextran by cariogenic bacteria and its presence in human dental plaque, *Arch. Oral Biol.* 12:11–24.

Gibbons, R. J., and Kapsimalis, B., 1963, Synthesis of intracellular iodophilic polysaccharide by *Streptococcus mitis, Arch. Oral Biol.* 8:319–329.

Gibbons, R. J., and Nyggaard, M., 1970, Interbacterial aggregation of plaque bacteria, *Arch. Oral Biol.* 15:1397–1400.

Gibbons, R. J., and Qureshi, J. V., 1976, Interactions of *Streptococcus mutans* and other oral bacteria with blood group reacting substances, in: *Microbial Aspects of Dental*

Caries (H. M. Stiles, W. J. Loesche, and T. C. O'Brien, eds.), *Microbiol. Abstr. Spec. Suppl.* 1:163–184.

Gibbons, R. J., and Socransky, S. S., 1962, Intracellular polysaccharide stored by organisms in dental plaques. Its relation to dental caries and microbiological ecology of the oral cavity, *Arch. Oral Biol.* 7:73–80.

Gibbons, R. J., and van Houte, J., 1973, On the formation of dental plaques, *J. Periodontol.* 44:347–360.

Gibbons, R. J., and van Houte, J., 1975a, Bacterial adherence in oral microbial ecology, *Annu. Rev. Microbiol.* 29:19–44.

Gibbons, R. J., and van Houte, J., 1975b, Dental caries, *Annu. Rev. Med.* 26:121–136.

Gibbons, R. J., Socransky, S. S., Sawyer, S., Kapsimalis, B., and MacDonald, J. B., 1963, The microbiota of the gingival crevice area of man. II. The predominant cultivable organisms, *Arch. Oral Biol.* 8:281–289.

Gibbons, R. J., Socransky, S. S., De Aravjo, W. C., and van Houte, J., 1964, Studies on the predominant cultivable flora of dental plaque, *Arch. Oral Biol.* 9:365–370.

Gibbons, R. J., De Paola, R. P., Spinell, D. M., and Skobe, Z., 1974, Interdental localization of *Streptococcus mutans* as related to dental caries experience, *Infect. Immun.* 9:481–488.

Gilmour, M. N., and Nisengard, R. J., 1974, Interactions between serum titres to filamentous bacteria and their relationship to human periodontal disease, *Arch. Oral Biol.* 19:959–968.

Gilmour, M. N., and Poole, A. E., 1967, The fermentation capacities of dental plaque, *Caries Res.* 1:247–267.

Gilmour, M. N., Green, G. C., Zahn, L. M., Sparmann, C. D., and Pearlman, J., 1976, The C_1–C_4 monocarboxylic and lactic acids in dental plaques before and after exposure to sucrose *in vivo*, in: *Microbial Aspects of Dental Caries* (H. M. Stiles, W. J. Loesche, and T. C. O'Brien, eds.), *Microbiol. Abstr. Spec. Suppl.* 2:539–556.

Gjermo, P., and Eriksen, H. M., 1974, Unchanged plaque inhibiting effect of chlorhexidine in human subjects after two years of continuous use, *Arch. Oral Biol.* 19:317–319.

Glantz, P. O., 1969, On wettability and adhesiveness, *Odontol. Revy* 20:1–132.

Gordon, D. F., Jr., and Gibbons, R. J., 1966, Studies on the predominant cultivable microorganisms from the human tongue, *Arch. Oral Biol.* 11:627–632.

Gordon, D. F., Jr., and Jong, B. B., 1968, Indigenous flora from human saliva, *Appl. Microbiol.* 16:428–429.

Gordon, H. A., and Pesti, L., 1971, The gnotobiotic animal as a tool in the study of host microbial relationships, *Bacteriol. Rev.* 35:390–429.

Guggenheim, B., 1970, Extracellular polysaccharides and microbial plaque, *Int. Dent. J.* 20:657–678.

Guggenheim, B., 1976, Ultrastructure and some biochemical aspects of dental plaque: A review, in: *Microbial Aspects of Dental Caries* (H. M. Stiles, W. J. Loesche, and T. C. O'Brien, eds.), *Microbiol. Abstr. Spec. Suppl.* 1:89–108.

Guggenheim, B., and Burckhardt, J. J., 1974, Isolation and properties of a dextranase from *Streptococcus mutans OMZ* 176, *Helv. Odontol. Acta* 18:101–113.

Guggenheim, B., Muhlerman, H. R., Regolati, B., and Schmid, R., 1970, The effect of immunization against streptococci on glucosyl transferases on plaque formation and dental caries in rats, in: *Dental Plaque* (W. D. McHugh, ed.), pp. 287–296, E. and S. Livingstone, London.

Hadi, A. W., and Russell, C., 1968, Quantitative estimations of fusiforms in saliva from normal individuals and cases of acute ulcerative gingivitis, *Arch. Oral Biol.* 13:1371–1376.

Hamilton, I. R., 1968, Synthesis and degradation of intracellular polyglucose in *Streptococcus salivarius*, *Can. J. Microbiol.* 14:67–77.

Hamilton, I. R., 1969, Growth characteristics of adapted and ultraviolet induced mutants of *Streptococcus salivarius* resistant to sodium fluoride, *Can. J. Microbiol.* 15:287–295.

Hamilton, I. R., 1976, Intracellular polysaccharide synthesis by cariogenic microorganisms, in: *Microbial Aspects of Dental Caries* (H. M. Stiles, W. J. Loesche, and T. C. O'Brien, eds.), *Microbiol. Abstr. Spec. Suppl.* 3:683–701.

Hamilton, I. R., 1977, Effects of fluoride on enzymatic regulation of bacterial carbohydrate metabolism, *Caries Res.* 11:262–291.

Hamilton, I. R., and Ellwood, D. C., 1977, PEP phosphotransferase activity in *Actinomyces viscosus*. Evidence for fluoride resistance, *J. Dent. Res.* 56:310.

Hamilton, I. R., and Ellwood, D. C., 1978a, Effects of fluoride on carbohydrate metabolism by washed cells of *Streptococcus mutans* grown at various pH values in a chemostat, *Infect. Immun.* 19:434–442.

Hamilton, I. R., and Ellwood, D. C., 1978b, Fluoride sensitivity and phosphotransferase (PT) activity in cells of *Streptococcus mutans* Ingbritt grown in the chemostat, *J. Dent. Res.* 57:800.

Hammond, B. F., 1971, Intracellular polysaccharide production by human oral strains of *Lactobacillus casei*, *Arch. Oral Biol.* 16:323–338.

Hammond, B. F., and Darkes, M., 1976, Plasmids of *Lactobacillus casei*, in: *Microbial Aspects of Dental Caries* (H. M. Stiles, W. J. Loesche, and T. C. O'Brien, eds.), *Microbiol. Abstr. Spec. Suppl.* 3:737–747.

Handleman, S. L., and Mills, J. R., 1965, Enumeration of selected salivary bacterial groups, *J. Dent. Res.* 44:1343–1353.

Hardie, J. M., and Bowden, G. H., 1974a, The normal microbial flora of the mouth, in: *The Normal Microbial Flora of Man* (G. Sykes and F. A. Skinner, eds.), pp. 47–83, Academic Press, London.

Hardie, J. M., and Bowden, G. H., 1974b, Cell wall and serological studies on *Streptococcus mutans*, *Caries Res.* 8:301–316.

Hardie, J. M., and Bowden, G. H., 1976a, The microbial flora of dental plaque: Bacterial succession and isolation considerations, in: *Microbial Aspects of Dental Caries* (H. M. Stiles, W. J. Loesche, and T. C. O'Brien, eds.), *Microbiol. Abstr. Spec. Suppl.* 1:63–87.

Hardie, J. M., and Bowden, G. H., 1976b, Physiological classification of oral viridans streptococci, *J. Dent. Res. Spec. Iss. A* 55:166–176.

Hardie, J. M., and Marsh, P. J., 1979, Streptococci and the human oral flora, in: *Streptococci* (F. Skinner and L. Quesnel, eds.), Academic Press, London (in press).

Hardie, J. M., Thomson, P. L., South, R. J., Marsh, P. D., Bowden, G. H., McKee, A. S., Fillery, E. D., and Slack, G. L., 1977, A longitudinal epidemiological study on dental plaque and the development of dental caries, *J. Dent. Res. Spec. Iss. C* 56:90–98.

Hay, D. I., and Hartles, R. L., 1965, The effect of saliva on the metabolism of the oral flora, *Arch. Oral Biol.* 10:485–498.

Hayes, M. L., and Hyatt, A. L., 1974, The decarboxylation of amino acids by bacteria derived from human dental plaque, *Arch. Oral Biol.* 19:361–369.

Herremans, J. F., 1974, Immunoglobin A, in: *The Antigens* (M. Sela, ed.), Vol. II, pp. 365–522, Academic Press, New York.

Hofstad, T., 1974, Antibodies reacting with lipopolysaccharides from *Bacteroides melaninogenicus, Bacteroides fragilis,* and *Fusobacterium nucleatum* in serum from normal human subjects, *J. Infect. Dis.* 129:349–352.

Hofstad, T., and Skaug, N., 1978, A polysaccharide antigen from the Gram positive organism *Eubacterium saburreum* containing dideoxyhexose as the immunodominant sugar, *J. Gen. Microbiol.* 106:227–232.

Holbrook, I. B., and Molan, P. C., 1975, The identification of a peptide in human parotid saliva particularly active in enhancing the glycolytic activity of the salivary microorganisms, *Biochem. J.* 149:489–492.

Holbrook, W. P., and Duerden, B. I., 1974, A comparison of some characteristics of reference strains of *Bacteroides oralis* with *Bacteroides melaninogenicus, Arch. Oral Biol.* 19:1231–1235.

Holdeman, L. V., and Moore, W. E. C., 1977, *Anaerobe Laboratory Manual,* Anaerobe Laboratory, Virginia Polytechnic Institute and State University, Blacksburg, Virginia.

Holmberg, K., 1976, Isolation and identification of Gram-positive rods in human dental plaque, *Arch. Oral Biol.* 21:153–160.

Holmberg, K., and Hallander, H. O., 1973, Production of bactericidal concentrations of hydrogen peroxide by *Streptococcus sanguis, Arch. Oral Biol.* 18:423–434.

Holmberg, K., and Nord, C. E., 1975, Numerical taxonomy and laboratory identification of *Actinomyces* and *Arachnia* and some related bacteria, *J. Gen. Microbiol.* 91:1744–1751.

Holten, E., 1974, Glucokinase and glucose-6-phosphate dehydrogenase in *Neisseria, Acta Pathol. Microbiol. Scand. Sect. B* 82:201–206.

Hoogendoorn, H., 1976, The inhibitory action of the lactoperioxidase system on *Streptococcus mutans* and other microorganisms, in: *Microbial Aspects of Dental Caries* (H. M. Stiles, W. J. Loesche, and T. C. O'Brien, eds.), *Microbiol. Abstr. Spec. Suppl.* 2:353–357.

Hoshino, E., Yamada, T., and Araya, S., 1976, Lactate degradation by a strain of *Neisseria* isolated from human dental plaque, *Arch. Oral Biol.* 21:677–683.

Hotz, P., Guggenheim, B., and Schmid, R., 1972, Carbohydrates in pooled dental plaque, *Caries Res.* 6:103–121.

Howell, A., and Jordan, H. V., 1963, A filamentous microorganism isolated from periodontal plaque in hamsters. II. Physiological and biochemical characteristics, *Sabouraudia* 3:93–105.

Howell, A., Jr., Rizzo, A., and Paul, F., 1965, Cultivable bacteria in developing and mature human dental calculus, *Arch. Oral Biol.* 10:307–313.

Hughes, R. C., 1975, The complex carbohydrates of mammalian cell surfaces and their biological roles, *Essays Biochem.* 11:1–38.

Huis, int'Veld, J. H. J., van Palenstein Helderman, W. H., and Sampaio Camargo, P., 1978, A sequential study of the incidence and prevalence of *Streptococcus mutans* serotypes on approximal surfaces and in fissures, *J. Dent. Res. Spec. Iss. A* 57:142, Abstract No. 272.

Hunter, J. R., Baird, J. K., and Ellwood, D. C., 1973, Effect of fluoride on the transport of sugars into chemostat grown *Streptococcus mutans, J. Dent. Res.* 52:954.

Hurtado, R. C., Rola-Pleszczynski, M., Merida, M. A., Hensen, S. A., Vincent, M. M., Thong, Y. H., and Bellanti, J. A., 1975, The immunologic role of tonsillar tissues in local cell mediated immune responses, *J. Pediatr.* 86:405–408.

Hutchinson, G. E., 1965, The niche: An abstractly inhabited hyper-volume, in: *The Ecological Theater and the Evolutionary Play,* Chapter 2, Yale University Press, New Haven, Connecticut.

Huxley, H. G., 1972, The recovery of microorganisms from the fissures of rat molar teeth, *Arch. Oral Biol.* 17:1481–1485.

Huxley, H. G., 1976, The relationship between plaque bacteria and dental caries at specific tooth sites in rats, in: *Microbial Aspects of Dental Caries* (H. M. Stiles, W. J. Loesche, and T. C. O'Brien, eds.), *Microbiol. Abstr. Spec. Suppl.* 3:773-784.

Ikeda, T., Sandham, H. J., and Bradley, E. L., 1973, Changes in *Streptococcus mutans* and lactobacilli in relation to the initiation of dental caries in Negro children, *Arch. Oral Biol.* 18:555-566.

Imfeld, T., 1977, Evaluation of the cariogenicity of confectionary by intra-oral wire-telemetry, *Helv. Odontol. Acta* 21:1-28.

Ivanyi, L., and Lehner, T., 1970, Stimulation of lymphocyte transformation by bacterial antigens in patients with periodontal disease, *Arch. Oral Biol.* 15:1089-1096.

Iwani, Y., Higuchi, M., Yamada, T., and Araya, S., 1972, Degradation of lactate by *Bacterionema matruchotii* under aerobic and anaerobic conditions, *J. Dent. Res.* 51: 1683.

Jantzen, E., Bryn, K., Bergan, T., and Boure, K., 1975, Gas chromatography of bacterial whole cell methanolysates, *Acta Pathol. Microbiol. Scand.* 83:569-580.

Jenkins, G. N., Edgar, W. M., and Ferguson, D. B., 1969, The distribution and metabolic effects of human plaque fluorine, *Arch. Oral Biol.* 14:105-119.

Johnson, D. A., Behling, U. H., Lai, C. H., Listgarten, M., Socransky, S., and Nowotny, A., 1978, Role of bacterial products in periodontitis: Immune response in gnotobiotic rats monoinfected with *Eikenella corrodens*, *Infect. Immun.* 19:905-913.

Johnson, M. C., Bozzola, J. J. and Shechmeister, I. L., 1974, Morphological study of *Streptococcus mutans* and two extracellular polysaccharide mutants, *J. Bacteriol.* 118: 304-311.

Johnson, R., and Sneath, P. H. A., 1973, Taxonomy of *Bordetella* and related organisms of the families Achromobacteriaceae, Brucellaceae and Neisseriaceae, *Int. J. Syst. Bacteriol.* 23:381-404.

Jordan, H. V., 1965, Bacteriological aspects of experimental dental caries, *Ann. N.Y. Acad. Sci.* 131:905-913.

Jordan, H. V., and Keyes, P. H., 1964, Aerobic, gram positive, filamentous bacteria as etiologic agents of experimental periodontal disease in hamsters, *Arch. Oral Biol.* 9: 401-414.

Jordan, H. V., and Sumney, D. L., 1973, Root surface caries: Review of the literature and significance of the problem, *J. Periodontol.* 44:158-163.

Jordan, H. V., Englander, H. R., and Lim, S., 1969, Potentially cariogenic streptococci in selected population groups in the western hemisphere, *J. Am. Dent. Assoc.* 78:1331-1335.

Jordan, H. V., Keyes, P. H., and Bellack, S., 1972, Periodontal lesions in hamsters and gnotobiotic rats infected with *Actinomyces* of human orgin, *J. Periodontol. Res.* 7: 21-28.

Kanapka, J. A., and Hamilton, I. R., 1971, Fluoride inhibition of enolase activity *in vivo* and its relationship to the inhibition of glucose-6-P formation in *Streptococcus salivarius*, *Arch. Biochem. Biophys.* 146:167-174.

Kanapka, J. A., Khandelwal, R. L., and Hamilton, I. R., 1971, Fluoride inhibition of glucose-6-P formation in *Streptococcus salivarius*: Relation to glycogen synthesis and degradation, *Arch. Biochem. Biophys.* 2:596-602.

Kaslick, R. S., Tuckman, M. A., and Chasens, A. I., 1973, Effect of topical vancomycin on plaque and chronic gingival inflammation, *J. Periodontol.* 44:366-368.

Kaufman, E. J., Mashimo, P. A., Hausmann, E., Hanks, C. T., and Ellison, S. A., 1972, Fusobacterial infection: Enhancement by cell free extracts of *Bacteroides melaninogenicus* possessing collagenolytic activity, *Arch. Oral Biol.* 17:577-580.

Kelstrup, J., and Gibbons, R. J., 1969, Bacteriocins from human and rodent streptococci, *Arch. Oral Biol.* 14:251–258.

Kelstrup, J., Richmond, S., West, C., and Gibbons, R. J., 1971, Fingerprinting human oral streptococci by bacteriocin production and sensitivity, *Arch. Oral Biol.* 15:1109–1116.

Kenney, E. B., and Ash, M. M., Jr., 1969, Oxidation–reduction potential of developing plaque, periodontal pockets and gingival sulci, *J. Periodontol.* 40:630–633.

Keyes, P. H., 1968, Research in dental caries, *J. Am. Dent. Assoc.* 76:1357–1373.

Khandelwal, R. L., Spearman, T. N., and Hamilton, I. R., 1972, Isolation and characterization of glycogen from *Streptococcus salivarius, Can. J. Biochem.* 50:140–142.

Killian, M., 1976, A taxonomic study of the genus *Haemophilus* with the proposal of a new species, *J. Gen. Microbiol.* 93:9–62.

Killian, W. F., and Ennever, J., 1975, Effect of magnesium on *Bacterionema matruchotii* calcification, *J. Dent. Res.* 54:185.

Kleinberg, I., 1961, Studies on dental plaque. I. The effect of different concentrations of glucose on the pH of dental plaque *in vivo, J. Dent. Res.* 40:1087–1111.

Kleinberg, I., 1967a, Effect of varying sediment and glucose concentrations on the pH and production in human salivary sediment mixtures, *Arch. Oral Biol.* 12:1457–1473.

Kleinberg, I., 1967b, Effect of urea concentration on human plaque pH levels *in situ, Arch. Oral Biol.* 12:1475–1484.

Kleinberg, I., 1970, Regulation of the acid–base metabolism of the dento-gingival plaque and its relation to dental caries and periodontal disease, *Int. Dent. J.* 29:451–465.

Kleinberg, I., and Jenkins, G. N., 1964, The pH of dental plaques in the different areas of the mouth before and after meals and their relationship to the pH and rate of flow of resting saliva, *Arch. Oral Biol.* 9:493–516.

Kleinberg, I., Craw, D., and Komiyama, K., 1973, Effect of salivary supernatant on the glycolytic activity of the bacteria in salivary sediment, *Arch. Oral Biol.* 18:787–798.

Kleinberg, I., Kanapka, J. A., and Craw, D., 1976, Effect of saliva and salivary factors on the metabolism of the mixed oral flora, in: *Microbial Aspects of Dental Caries* (H. M. Stiles, W. J. Loesche, T. C. O'Brien, eds.), *Microbiol. Abstr. Spec. Suppl.* 2:433–464.

Kligler, I. J., 1915, A biochemical study and differentiation of oral bacteria with special reference to dental caries, *J. Allied Dent. Soc.* 10:445–458.

Koch, G. (ed.), 1977, *International Working Seminar on Surface and Colloid Phenomena in the Oral Cavity, Swed. Dent. J.* 1:205–272.

Kocur, M., Bergen, T., and Mortensen, N., 1971, DNA base composition of Gram-positive cocci, *J. Gen. Microbiol.* 69:167.

Kohler, B., and Bratthall, D., 1976, Intrafamilial levels of *Streptococcus mutans*; a comparison between mother and child, *J. Dent. Res. Spec. Iss. A* 56: Abstract No. 457.

Krasse, B., 1968, Effects of dietaries on oral microbiology, in: *Art and Science of Dental Caries Research* (R. S. Harris, ed.), pp. 111–124, Academic Press, New York.

Krasse, B., and Carlsson, J., 1970, Various types of streptococci and experimental caries in hamsters, *Arch. Oral Biol.* 15:25–32.

Krasse, B., and Jordan, H. V., 1977, Effect of orally applied vaccines on oral colonization by *Streptococcus mutans* in rodents, *Arch. Oral Biol.* 22:479–484.

Krasse, B., Edwardson, S., Svenson, I., and Trell, L., 1967, Implantation of caries-inducing streptococci in the human oral cavity, *Arch. Oral Biol.* 12:231–236.

Krasse, B., Jordan, H. V., and Edwardson, S., 1968, The occurrence of certain "caries-inducing" streptococci in human dental plaque material with special reference to the frequency and activity of caries, *Arch. Oral Biol.* 13:911–918.

Kristoffersen, T., and Hofstad, T., 1970, Antibodies in humans to an isolated antigen from oral fusobacteria, *J. Periodontal Res.* 5:110–115.

Lang, N. P., and Smith, F. N., 1977, Lymphocyte blastogenesis to plaque antigens in human periodontal disease. I. Populations of varying severity of disease, *J. Periodontal Res.* 12:298–309.

Larje, O., and Frostell, G., 1968, Acid production activities of caries-inducing streptococci, *Acta Pathol. Microbiol. Scand.* 72:463.

Leach, S. A., and Hayes, M. L., 1968, A possible correlation between the specific bacterial enzyme activities, dental plaque formation and cariogenicity, *Caries Res.* 2:38–46.

Leach, S. A., and Melville, T., 1970, Investigation of some human oral organisms capable of releasing the carbohydrate from salivary glycoproteins, *Arch. Oral Biol.* 15:87–88.

Leach, S. A., Appleton, J., Dada, O. A., and Hayes, M. L., 1972, Some factors affecting the metabolism of fructan by human oral flora, *Arch. Oral Biol.* 17:137–146.

Lehner, T. (ed.), 1977a, *The Borderland between Caries and Periodontal Disease,* Academic Press, London.

Lehner, T., 1977b, Immunological mechanism in caries and gingivitis, in: *The Borderland between Caries and Periodontal Disease* (T. Lehner, ed.), pp. 129–144, Academic Press, London.

Lehner, T., Challacombe, S. J., and Caldwell, J., 1975a, An experimental model for immunological studies of dental caries in the rhesus monkey, *Arch. Oral Biol.* 20:299–304.

Lehner, T., Challacombe, S. J., and Caldwell, J., 1975b, An immunological investigation into the prevention of caries in deciduous teeth of rhesus monkeys, *Arch. Oral Biol.* 20:305–310.

Lehner, T., Challacombe, S. J., Wilton, J. M. A., and Caldwell, J., 1976a, Cellular and humoral immune responses in vaccination against dental caries in monkeys, *Nature (London)* 264:69–71.

Lehner, T., Challacombe, S. J., Wilton, J. M. A. and Ivanyi, L. 1976b, Immunoprotentiation by dental microbial plaque and its relationship to oral diseases in man, *Arch. Oral Biol.* 21:749–754.

Levine, M. J., Herzberg, M. C., Levine, M. S., Ellison, S. A., Stinson, M. W., Li, H. C., and van Dyke, T., 1978, Specificity of salivary–bacterial interactions: Role of terminal sialic acid residues in the interaction of salivary glycoproteins with *Streptococcus sanguis* and *Streptococcus mutans, Infect. Immun.* 19:107–116.

Liljemark, W. F., and Gibbons, R. J., 1971, Ability of *Veillonella* and *Neisseria* species to attach to oral surfaces and their proportions present indigenously, *Infect. Immun.* 4:264–268.

Liljemark, W. F., and Gibbons, R. J., 1973, Suppression of *Candida albicans* by human oral streptococci in gnotobiotic mice, *Infect. Immun.* 8:846–849.

Listgarten, M. A., 1976, Structure of the microbial flora associated with periodontal disease and health in man. A light and electron microscope study, *J. Periodontol.* 47:1–18.

Listgarten, M. A., and Lewis, D. W., 1967, The distribution of spirochetes in the lesion of acute necrotising ulcerative gingivitis: An electron microscopic and statistical survey, *J. Periodontol.* 38:379–386.

Listgarten, M. A., Mayo, H., and Amsterdam, M., 1973, Ultrastructure of the attachment device between coccal and filamentous microorganisms in "corn cob" formations of dental plaque, *Arch. Oral Biol.* 18:651–656.

Listgarten, M. A., Mayo, H. E., and Tremblay, R., 1975, Development of dental plaque on epoxy resin crowns in man, *J. Periodontol.* 46:10–26.

Littleton, N. W., and White, C. F., 1964, Dental findings from a preliminary study of children receiving extended antibiotic therapy, *J. Am. Dent. Assoc.* 68:520–526.

Littleton, N. W., McCabe, R. M, and Carter, C. H., 1967, Studies of oral health in persons nourished by stomach tube II. Acidogenic properties and selected bacterial components of plaque material, *Arch. Oral Biol.* 12:601–609.

Llory, H., Guillo, B., and Frank, R. M., 1971, A cariogenic *Actinomyces viscosus*. A bacteriological and gnotobiotic study, *Helv. Odontol. Acta* 15:134–138.

Löe, H., and Schiøtt, C. R., 1970, The effect of mouthrinses and topical application of chlorhexidine on the development of dental plaque and gingivitis in man, *J. Periodontol.* 3:79–83.

Löe, H., and Silness, J., 1963, Periodontal disease in pregnancy. I. Prevalence and severity, *Acta Odontol. Scand.* 21:533–551.

Löe, H., Theilade, E., and Jensen, S. B., 1965, Experimental gingivitis in man, *J. Periodontol.* 36:177–187.

Löe, H., Karring, T., and Theilade, E., 1973, An *"in vivo"* method for the study of the microbiology of occlusal fissures, *Caries Res.* 7:120–129.

Löe, H., Schiøtt, C. R., Glavind, L., and Karring, T., 1976, Two years oral use of chlorhexidine in man I. General design and clinical effects, *J. Periodontal Res.* 11:135–144.

Loesche, W. J., 1968, Importance of nutrition in gingival crevice microbial ecology, *Periodontics* 6:245–249.

Loesche, W. J., 1975, Bacterial succession in dental plaque: Role in dental disease, in: *Microbiology—1975* (D. E. Schlessinger, ed.), pp. 132–136, American Society for Microbiology, Washington, D.C.

Loesche, W. J., 1976a, Chemotherapy of dental plaque infections, *Oral Sci. Rev.* 9:65–107.

Loesche, W. J., 1976b, Periodontal disease and the treponemes, in: *The Biology of Parasitic Spirochetes* (R. C. Johnson, ed.), pp. 261–275, Academic Press, New York.

Loesche, W. J., 1977, Topical fluorides as an antibacterial agent, *J. Prev. Dent.* 4:21–26.

Loesche, W. J., and Gibbons, R. J., 1968, Amino acid fermentation by *Fusobacterium nucleatum*, *Arch. Oral Biol.* 13:191–201.

Loesche, W. J., and Henry, C. A., 1967, Intracellular microbial polysaccharide production and dental caries in a Guatemalan Indian village, *Arch. Oral Biol.* 12:189–194.

Loesche, W. J., and Syed, S. A., 1973, The predominant cultivable flora of carious plaque and carious dentine, *Caries Res.* 7:201–216.

Loesche, W. J., and Syed, S. A., 1978, Bacteriology of human experimental gingivitis: Effect of plaque and gingivitis score, *Infect. Immun.* 21:830–839.

Loesche, W. J., Hockett, R. N., and Syed, S. A., 1972, The predominant cultivable flora of tooth surface plaque removed from institutionalized subjects, *Arch. Oral Biol.* 17:1311–1325.

Loesche, W. J., Rowan, J., Straffon, L. H., and Loos, P. J., 1975a, Association of *Streptococcus mutans* with human dental decay, *Infect. Immun.* 11:1252–1260.

Loesche, W. J., Syed, S. A., Murray, R. J., and Mellberg, J. R., 1975b, Effect of topical acidulated phosphate fluoride on percentage of *Streptococcus mutans* and *Streptococcus sanguis* in plaque, *Caries Res.* 9:139–155.

Loesche, W. J., Bradbury, D. R., and Woolfolk, M. P., 1977, Reduction of dental decay in rampant caries individuals following short-term kanamycin treatment, *J. Dent. Res.* 56:254–265.

Loesche, W. J., Straffon, L. H., and Walker, M. C., 1978, Interim report on longitudinal studies of *Streptococcus mutans* colinization of human caries free fissures, *J. Dent. Res. Spec. Iss. A* 57:Abstract No. 790.

Long, S. S., and Svenson, R. M., 1976, Determinants of the developing oral flora in normal newborns, *Appl. Environ. Microbiol.* 32:494–497.

MacDonald, J. B., and Gibbons, R. J., 1962, The relationship of indigenous bacteria to periodontal disease, *J. Dent. Res.* 41:320–326.

MacDonald, J. B., Sutton, R. M., and Knoll, M. L., 1954, The production of fusospirochetal infections in guinea pigs with recombined pure cultures, *J. Infect. Dis.* 92:275–284.

MacDonald, J. B., Gibbons, R. J., and Socransky, S. S., 1960, Bacterial mechanisms in periodontal disease, *Ann. N.Y. Acad. Sci.* 85:467–478.

MacDonald, J. B., Socransky, S. S., and Gibbons, R. J., 1963, Aspects of the pathogenesis of mixed anaerobic infections of mucous membranes, *Bull. Tokyo Dent. Coll.* 18:217–229.

Maeder, C. L., Karge, H. J., Angel, I., and Newman, M. G., 1978, Comparison of oral gliding bacteria isolated from various periodontal conditions with the order *Cytophagales*, *J. Dent. Res. Spec. Iss. A.* 57: Abstract No. 1105.

Makinen, K. K., and Scheinin, A., 1972, The effect of various sugars and sugar mixtures on the activity and formation of enzymes of dental plaque and oral fluid, *Acta Odontol. Scand.* 30:259–275.

Mandel, I. D., 1976, Salivary products in plaque and saliva in relation to caries, in: *Microbial Aspects of Dental Caries* (H. M. Stiles, W. J. Loesche, and T. C. O'Brien, eds.), *Microbiol. Abst. Spec. Suppl.* 3:859–866.

Mandel, I. D., 1978, Salivary factors in caries prediction, in: *Methods of Caries Prediction* (Workshop Proceedings) (B. G. Bibby, and R. J. Shern, eds.), *Microbiol. Abstr. Spec. Suppl.* pp. 147–158, Information Retrieval, Inc., Washington, D.C.

Manly, R. S., and Richardson, D. T., 1968, Metabolism of levan by oral samples, *J. Dent. Res.* 47:1080–1086.

Marsh, P. D., Bowden, G. H., and Hardie, J. M., 1978, Numerical fluctuations of *Streptococcus mutans* in approximal plaque, *J. Dent. Res.* (British Division IADR) 57:88.

Marshall, K. C., 1976, *Interfaces in Microbial Ecology*, Harvard University Press, Cambridge, Massachusetts.

Marthaler, T. M., and Germann, M., 1970, Radiographic and visual appearance of small smooth surface caries lesions studied on extracted teeth, *Caries Res.* 4:224–242.

Mattingly, S. J., Dipersio, T. R., Higgins, M. L., and Shockman, G. D., 1976, Unbalanced growth and macromolecules synthesis in *Streptococcus mutans FA-I*, *Infect. Immun.* 13:941–948.

Mattingly, S. J., Daneo-Moore, L., and Shockman, G. D., 1977, Factors regulating cell wall thickening and intracellular iodophilic polysaccharide storage in *Streptococscus mutans*, *Infect. Immun.* 16:967–973.

McCabe, R. M, and Donkersloot, J. A., 1977, Adherence of *Veillonella* species mediated by extracellular glucosyl-transferase from *Streptococcus salivarius*, *Infect. Immun.* 18:726–734.

McCarthy, C., Snyder, M. L., and Parker, P. B., 1965, The indigenous oral flora of man. I. The newborn to the 1 year old infant, *Arch. Oral Biol.* 10:61–70.

McGhee, J. R., Michalek, S. M., Webb, J., Nauia, J. M., Rahman, A. F. R., and Ledher, D. W., 1975, Effective immunity to dental caries: Protection of gnotobiotic rats by local immunization with *Streptococcus mutans*, *J. Immunol.* 114:300–305.

McGhee, J. R., Mestecky, J., and Babb, J. L. (eds), 1978, *Secretory Immunity and Infection*, Plenum Press, New York.

McHugh, W. D. (ed.), 1970, *Dental Plaque*, E. and S. Livingstone, Edinburgh.

McKee, A., and Shah, H. N., 1979, Identification of anaerobic Gram-negative rods from dental plaque, *J. Dent. Res.* (British Division IADR) 58:Abstract No. 165 (in press).

Melisch, D. F., Loesche, W. J., and Syed, S. A., 1978, Intrafamilial comparisons of bacteriocin codes of *Streptococcus mutans*, *J. Dent. Res. Spec. Iss. A* 57:Abstract No. 273.

Mestecky, J., 1976, Introduction to the structural and cellular aspects of the secretory IgA, system, *J. Dent. Res. Spec. Iss. C* 55:98–101.

Michalek, S. M., McGhee, J. R., Mestecky, J., Arnold, R. R., and Bozzo, L., 1976a, Ingestion of *Streptococcus mutans* induces secretory immunoglobulin A and caries immunity, *Science*, 192:1238–1240.

Michalek, S. M., McGhee, J. R., Nauia, J. M., and Narkates, A. J., 1976b, Effective immunity to dental caries: Protection of malnourished rats by local injection of *Streptococcus mutans*, *Infect. Immun.* 13:782–789.

Michalek, S. M., McGhee, J. R., and Babb, J. L., 1978, Effective immunity to dental caries: Dose dependent studies on secretory immunity by oral administration of *Streptococcus mutans* to rats, *Infect. Immun.* 19:217–224.

Michaud, R. N., and Delwiche, E. A., 1970, Multiple impairment of glycolysis of *Veillonella alcalescens*, *J. Bacteriol.* 101:138–140.

Michaud, R. N., Carrow, J. A., and Delwiche, E. A., 1970, Non-oxidative pentose phosphate pathway in *Veillonella alcalescens*, *J. Bacteriol.* 101:141–144.

Mikx, F. H. M., and Svanberg, M., 1978, Considerations about microbial interactions in relation to modification of the microflora of dental plaque, in: *Methods of Caries Prediction* (B. G. Bibby and R. J. Shern, eds.), *Microbiol. Abstr. Spec. Suppl.* pp. 109–118, Information Retrieval, Inc., Washington, D.C.

Mikx, F. H. M., and van der Hoeven, J. S., 1975, Symbiosis of *Streptococcus mutans* and *Veillonella alcalescens* in mixed continuous cultures, *Arch. Oral Biol.* 20:407–410.

Mikx, F. H. M., van der Hoeven, J. S., König, K. G., Plasschaert, A. J. M., and Guggenheim, B., 1972, Establishment of defined microbial ecosystems in germ-free rats, *Caries Res.* 6:211–223.

Mikx, F. H. M., van der Hoeven, J. S., Plasschaert, A. J. M., and König, K. G., 1975a, Effect of *Actinomyces viscosus* on the establishment and symbiosis of *Streptococcus mutans* and *Streptococcus sanguis* in SPF rats on different sucrose diets, *Caries Res.* 9:1–20.

Mikx, F. H. M., van der Hoeven, J. S., Plasschaert, A. J. M., and König, K. G., 1975b, Establishment and symbiosis of *Actinomyces viscosus*, *Streptococcus sanguis* and *Streptococcus mutans* in germ-free Osborne Mendel rats, *Caries Res.* 9:286–324.

Mikx, F. H. M., van der Hoeven, J. S., and Walker, G. J., 1976a, A microbial symbiosis in dental plaque studied in gnotobiotic rats and in the chemostat, in: *Microbial Aspects of Dental Caries* (H. M. Stiles, W. J. Loesche, and T. C. O'Brien, eds.), *Microbiol. Abstr. Spec. Suppl.* 3:763–772.

Mikx, F. H. M., van der Hoeven, J. S., Plasschaert, A. J. M., and Maltha, J. C., 1976b, Establishment of defined microbial ecosystems in germ-free rats. II. Diassociation of Osborne–Mendel rats with *Veillonella alcalescens* and several oral microorganisms, *Caries Res.* 10:49–58.

Miller, B. F., and Muntz, J. A., 1939, The quantity of lactic acid in carious areas of human teeth, and its relationship to the soluble calcium and phosphate in these lesions, *J. Dent. Res.* 18:259–265.

Miller, C. E., Wong, K. H., Feeley, J. C., and Forlies, M. E., 1972, Immunological conversion of *Vibrio cholerae* in gnotobiotic mice, *Infect. Immun.* 6:739–742.

Miller, C. H., and Kleinman, J. L., 1974, Effect of microbial interactions on *in vitro* plaque formation by *Streptococcus mutans*, *J. Dent. Res.* 53:427–434.

Miller, W. D., 1889, *Mikroorganismen der Mundhöhle*, Georg Thieme, Leipzig.

Minah, G. E., and Bowman, G., 1978, Identity and cariogenicity of the microflora colonizing removable enamel surfaces, *J. Dent. Res. Spec. Iss. A* 57:Abstract No. 426.

Minah, G. E., and Loesche, W. J., 1977a, Sucrose metabolism in resting cell suspensions of caries-associated and non-caries-associated dental plaque, *Infect. Immun.* 17:43–54.

Minah, G. E., and Loesche, W. J., 1977b, Sucrose metabolism by prominent members of the flora isolated from cariogenic and non-cariogenic dental plaques, *Infect. Immun.* 17:55–61.

Mishiro, Y., Kirimura, K., and Ishihara, H., 1966, Observations on the saliva factor that enhances the production of lactic acid by the oral flora, *J. Dent. Res.* 45:1824.

Mitchell, R., 1976, Mechanism of attachment of microorganisms to surfaces, in: *Microbial Aspects of Dental Caries* (H. M. Stiles, W. J. Loesche, and T. C. O'Brien, eds. *Microbiol. Abstr. Spec. Suppl.* 1:47–53.

Moore, B. W., Carter, W. J., Dunn, J. K., and Fosdick, L. S., 1956, The formation of lactic acid in dental plaques. I. Caries-active individuals, *J. Dent. Res.* 35:778–785.

Mühlemann, H. R., and de Boever, J., 1970, Radiotelemetry of the pH of interdental areas exposed to various carbohydrates, in: *Dental Plaque* (W. D. McHugh, ed.), pp. 179–186, E. and S. Livingstone, Edinburgh.

Muntz, J. A., 1943, Production of acids from glucose by dental plaque material, *J. Biol. Chem.* 148:225–236.

Naeslund, C., 1926, Studies of tartar formation, *Acta Pathol. Microbiol. Scand.* 3:637–677.

Nakamura, T., Suginaka, Y., Orata, T., Obata, N., and Yamazaki, N., 1977, Bacteriocin-like activities of human dental plaque flora against oral anaerobic microorganisms, *Bull. Tokyo Dent. Coll.* 18:217–229.

Neff, D.. 1967, Acid production from different carbohydrate sources in human plaque *in situ*, *Caries Res.* 1:78–87.

Newbrun, E., 1972, Water fluoridation and dietary fluoride, in: *Fluorides and Dental Caries* (E. Newbrun, ed.), Charles C. Thomas, Springfield, Illinois.

Newbrun, E., 1976, Polysaccharide synthesis in plaque, in: *Microbial Aspects of Dental Caries* (H. M. Stiles, W. J. Loesche, and T. C. O'Brien, eds.), *Microbiol. Abstr. Spec. Suppl.* 3:649–664.

Newman, H. N., and Poole, D. F. G., 1974, Structural and ecological aspects of dental plaque, in: *The Normal Microbial Flora of Man* (F. A. Skinner and J. G. Carr, eds.), *Soc. Appl. Bacteriol. Symp. Ser.*, No. 3:111–134.

Newman, M. G., and Socransky, S. S., 1977, Predominant cultivable microbiota in periodontosis, *J. Periodontal Res.* 12:120–128.

Newman, M. G., Williams, R. C., Crawford, A., Manganiello, A. D., and Socransky, S. S., 1973, Predominant cultivable microbiota of periodontitis and periodontosis. III. Periodontosis, *J. Dent. Res. Spec. Iss.* 52:Abstract No. 290.

Newman, M. G.. Socransky, S. S.. Sarritt, E. D., Propas, D. A.. and Crawford, A., 1976. Studies of the microbiology of periodontosis, *J. Periodontol.* 47:373–379.

Ng, S. K. C., and Hamilton, I. R., 1971, Lactate metabolism by *Veillonella parvula*, *J. Bacteriol.* 105:999–1005.

Ng, S. K. C., and Hamilton, I. R., 1973, Carbon dioxide fixation by *Veillonella parvula* M4 and its relation to propionic acid formation, *Can. J. Microbiol.* 19:715–723.

Ng, S. K. C., and Hamilton, I. R., 1974, Gluconeogensis by *Veillonella parvula* M4: Evidence for the indirect conversion of pyruvate to *P*-enolpyruvate, *Can. J. Microbiol.* 20:19–28.

Ng, S. K. C., and Hamilton, I. R., 1975, Purification and regulatory properties of pyruvate kinase from *Veillonella parvula*, *J. Bacteriol.* 122:1274–1282.

Nisengard, R. J., 1977, The role of immunology in periodontal disease, *J. Periodontal.* 48:505–515.

Orstavik, D., and Brandtzaeg, P., 1977, Serum antibodies to plaque bacteria in subjects with dental caries and gingivitis, *Scand. J. Dent. Res.* 85:106–113.

Palenik, C. J. and Miller, C. H., 1975, Extracellular invertase activity from *Actinomyces viscosus*, *J. Dent. Res.* 54:186.

Parker, R. B., 1970, Paired culture interaction of the oral microbiota, *J. Dent. Res.* 49:804–809.

Parker, R. B., and Creamer, H. R., 1971, Contribution of plaque polysaccharides to growth of cariogenic microorganisms, *Arch. Oral Biol.* 16:855–862.

Pine, L., and Howell, A., 1956, Comparison and biochemical characters of *Actinomyces* spp. with those of *Lactobacillus bifidus*, *J. Gen. Microbiol.* 15:428–435.

Posner, A. S., 1969, Crystal chemistry of bone mineral, *Physiol. Rev.* 49:760–792.

Raina, J. L., and Ravin, A. W., 1976, Enhanced transformability with heterospecific deoxyribonucleic acid in a *Streptococcus sanguis* mutant impaired in ribonucleic acid polymerase activity, *J. Bacteriol.* 127:380–391.

Ranney, R. R., 1970, Specific antibody in gingiva and submandibular nodes of monkeys with allergic periodontal disease, *J. Periodontal Res.* 5:1–7.

Renggli, H. H., 1977, Phagocytosis and killing by crevicular neutrophils, in: *The Borderland between Caries and Periodontal Disease* (T. Lehner, ed.), pp. 221–222, Academic Press, London.

Ritz, H. L., 1967, Microbial population shifts in developing human dental plaque, *Arch. Oral Biol.* 12:1561–1568.

Rizzo, A. A., and Mitchell, G. T., 1966, Chronic allergic inflammation induced by repeated deposition of antigen in rabbit gingival pockets, *Periodontics* 4:5–10.

Rizzo, A. A., Scott, D. B., and Bladen, H. A., 1963, Calcification of oral bacteria, *Ann. N.Y. Acad. Sci.* 109:14–22.

Robrish, S. A., and Krichevsky, M. I., 1972, Acid production from glucose and sucrose by growing cultures of caries-conducive streptococci, *J. Dent. Res.* 51:734–739.

Rogers, A. H., 1975, Bacteriocin types of *Streptococcus mutans* in human mouths, *Arch. Oral Biol.* 20:853–858.

Rogers, A. H., 1976a, Bacteriocin patterns of strains belonging to various serotypes of *Streptococcus mutans*, *Arch. Oral Biol.* 21:243–249.

Rogers, A. H., 1976b, Bacteriocinogeny and the properties of some bacteriocins of *Streptococcus mutans*, *Arch. Oral Biol.* 21:99–104.

Rogers, A. H., van der Hoeven, J. S., and Mikx, F. H. M., 1978, Inhibition of *Actinomyces viscosus* by bacteriocin-producing strains of *Streptococcus mutans* in the dental plaque of gnotobiotic rats, *Arch. Oral Biol.* 23:477–483.

Rogosa, M., 1964, The genus *Veillonella*, 1. General cultural, ecological and biochemical considerations, *J. Bacteriol.* 87:162–170.

Rogosa, M., 1965, The genus *Veillonella*. IV. Serological groupings, and genus and species emendations, *J. Bacteriol.* 90:704–709.

Rogosa, M., 1970, Characters used in the classification of lactobacilli, *Int. J. Syst. Bacteriol.* 20:519–533.

Rogosa, M., and Bishop, F. S., 1964, The genus *Veillonella*. 2. Nutritional studies. *J. Bacteriol.* 87:574–580.

Rogosa, M., Krichevsky, M. I., and Bishop, F. S., 1965, Truncated glycolytic system in *Veillonella*, *J. Bacteriol.* 90:164–171.

Rolla, G., 1976, Inhibition of adsorption—general considerations, in: *Microbial Aspects of Dental Caries* (H. M. Stiles, W. J. Loesche, and T. C. O'Brien, eds.), *Microbiol. Abstr. Spec. Suppl.* 2:309–324.

Rolla, G., and Melsen, B., 1975, Desorption of protein and bacteria from hydroxyapatite by fluoride and monofluorophosphate, *Caries Res.* 9:66–73.

Rolla, G., Löe, H., and Schiøtt, R., 1970, The affinity of chlorhexidine for hydroxyapatite and salivary mucins, *J. Periodontal Res.* 5:90–95.

Rosan, B., and Hammond, B. F., 1974, Extracellular polysaccharides of *Actinomyces viscosus*, *Infect. Immun.* 10:304–308.

Roseman, S., 1970, The synthesis of complex carbohydrates by multi-glycosyl transferase systems and their potential role in intercellular adhesion. *Chem. Phys. Lipids* 5:270–297.

Roth, S., McGuire, E. J., and Roseman, S., 1971, Evidence for cell surface glycosyl transferases, their potential role in cellular recognition, *J. Cell Biol.* 51:536–547.

Rugg-Gunn, A. J., Holloway, P. J., and Davies, T. G. H., 1973, Caries prevention by daily fluoride mouthrinsing, *Br. Dent. J.* 135:353–360.

Russell, A. L., 1956, A system for classification and scoring for prevalence surveys of periodontal disease, *J. Dent. Res.* 35:350–359.

Russell, R. R. B., and McDonald, I. J., 1976, Comparison of the cell envelope proteins of *Micrococcus cryophilus* with those of *Neisseria* and *Branhamella* species, *Can. J. Microbiol.* 22:309–312.

Russell, R. R. B., Johnson, K. G., and McDonald, I. J., 1975, Envelope proteins in *Neisseria*, *Can. J. Microbiol.* 21:1519–1534.

Rutter, P. R., 1979, The accumulation of organisms on the teeth, in: *Adhesion of Microorganisms to Surfaces* (D. C. Ellwood, J. Melling, and P. Rutter, eds.), Academic Press, London (in press).

Rutter, P. R., and Abbott, A., 1978, A study of the interaction between oral streptococci and hard surfaces, *J. Gen. Microbiol.* 105:219–226.

Sandham, H. J., and Kleinberg, I., 1969, The effect of fluoride on the interrelation between glucose utilization, pH and carbohydrate storage in a salivary sediment system, *Arch. Oral Biol.* 14:619–628.

Sanyal, B., and Russell, C., 1978, Non-sporing anaerobic Gram-positive rods in saliva at the gingival crevice of humans, *Appl. Environ. Microbiol.* 35:670–678.

Sasaki, S., Slots, J., Hammond, B., and Socransky, S. S., 1978, Enumeration of *Bacteroides melaninogenicus* and *Capnocytophaga* sp. in subgingival plaque by fluorescent antibody technique, *J. Dent. Res. Spec. Iss. A* 57:Abstract No. 966.

Savage, D. C., 1977, Interactions between the host and its microbes, in: *Microbial Ecology of the Gut* (R. T. J. Clarke and T. Bauchop, eds.), pp. 277–310, Academic Press, London.

Saxton, C. A., 1973, Scanning electron microscope study of the formation of dental plaque, *Caries Res.* 7:102–119.

Saxton, C. A., 1975a, The formation of dental plaque: A study by electron microscopy, Ph.D. Thesis, London University, London.

Saxton, C. A., 1975b, Determination by electron microscope autoradiography of the distribution in plaque of organisms that synthesize intracellular polysaccharide *in situ*, *Caries Res.* 9:418–437.

Scardovi, V., and Crociani, F., 1974, *Bifidobacterium catenulatium, Bifidobacterium dentium* and *Bifidobacterium angulatum*: Three new species and their deoxynucleic acid homology relationships, *Int. J. Syst. Bacteriol.* 24:6–20.

Schachtele, C. F., and Mayo, J. A., 1973, Phosphoenolpyruvate-dependent glucose transport in oral streptococci, *J. Dent. Res.* 52:1209–1215.

Schachtele, C. F., Staat, R. H., and Harlander, S. K., 1975, Dextranases of oral bacteria: Inhibition of water insoluble glucan production and adherence to smooth surfaces by *Streptococcus mutans, Infect. Immun.* 12:309–317.

Schamschula, R. G., Adkins, B. L., Barmes, D. E., Charlton, G., and Davey, B. G., 1978, WHO Study of Dental Caries Etiology in Papua, New Guinea, WHO Offset Publication No. 40, World Health Organization, Geneva.

Schiøtt, C. R., Briner, W. W., and Löe, H., 1976, Two years oral use of chlorhexidine in man. II. The effect on salivary bacteriol flora, *J. Periodontal Res.* 11:145–152.

Schroeder, H. E., 1969, *Formation and Inhibition of Dental Calculus*, Hans Huber, Berne, Switzerland.

Schroeder, H. E., 1977, Histopathology of the gingival sulcus, in: *The Borderland between Caries and Periodontal Disease* (T. Lehner, ed.), pp. 43–78, Academic Press, London.

Shah, H. N., Williams, R. A. D., Bowden, G. H., and Hardie, J. M., 1976, Comparison of the biochemical properties of *Bacteroides melaninogenicus* from human dental plaque and other sites, *J. Appl. Bacteriol.* **41**:473–492.

Sharma, M., Dhillon, A. S., and Newbrun, E., 1974, Cell-bound glucosyl transferase of *Streptococcus sanguis* strain 804, *Arch. Oral Biol.* **19**:1063–1073.

Shedlofsky, S., and Freter, R., 1974, Synergism between ecological and immunological control mechanisms of intestinal flora, *J. Infect. Dis.* **129**:296–303.

Shockman, G. D., Tsien, H. C., Kessler, R. E., Mychajlouka, M., Higgins, M. L., and Daneo-Moore, L., 1976, Effect of environmental factors on the surface properties of oral microorganisms, in: *Microbial Aspects of Dental Caries* (H. M. Stiles, W. J. Loesche, and T. C. O'Brien, eds.), *Microbiol. Abstr. Spec. Suppl.* **3**:631–648.

Sidaway, D. A., 1978a, A microbiological study of dental calculus. I. The microbial flora of mature calculus, *J. Periodontal Res.* **13**:349–359.

Sidaway, D. A., 1978b, A microbiological study of dental calculus. II. The *in vitro* calcification of microorganisms from dental calculus, *J. Periodontal Res.* **13**:360–366.

Silverstone, L. M., 1977, Remineralization phenomena, *Caries Res.* **11** (Suppl. 1): 59–84.

Sims, W., 1970, Oral haemophili, *J. Med. Microbiol.* **3**:615–625.

Sirisinha, S., 1970, Reaction of human salivary immunoglobulins with indigenous bacteria, *Arch. Oral Biol.* **15**:551–554.

Sirisinha, S., and Charupatana, C., 1971, Antibodies to indigenous bacteria in human serum, secretion and urine, *Can. J. Microbiol.* **17**:1471–1473.

Slack, G. L., and Bowden, G. H., 1965, Preliminary studies of experimental dental plaque "in vivo," *Adv. Fluorine Res. Dent. Caries Prev.* **3**:193–215.

Slack, J. M., and Gerencser, M. A., 1975, *Actinomyces, Filamentous Bacteria, Biology and Pathogenicity*, Burgess Publishing Co., Minneapolis, Minnesota.

Slots, J., 1976, The predominant cultivable organisms in juvenile periodontosis, *Scand. J. Dent. Res.* **84**:1–10.

Slots, J., 1977a, Microflora in the healthy gingival sulcus in man, *Scand. J. Dent. Res.* **85**:247–254.

Slots, J., 1977b, The predominant cultivable microflora of advanced periodontitis, *Scand J. Dent. Res.* **85**:114–121.

Slots, J., and Gibbons, R. J., 1978, Attachment of *Bacterioides melaninogenicus* subsp. *asaccharolyticus* to oral surfaces and its possible role in colonization of the mouth and periodontal pockets, *Infect. Immun.* **19**:254–264.

Slots, J., Moenbo, D., Largeback, J., and Frandsen, A., 1978, Microbiota of gingivitis in man, *Scand. J. Dent. Res.* **86**:174–181.

Smith, D. J., and Taubman, M. A., 1976, Immunization studies using the rodent caries model, *J. Dent. Res. Spec. Iss. C* **55**:193–205.

Smith, F. M., and Lang, N. P., 1977, Lymphocyte blastogenesis to plaque antigens in human periodontal disease. II. The relationship to clinical parameters, *J. Periodontal Res.* **12**:310–317.

Socranksy, S. S., 1970, Relationship of bacteria to the etiology of periodontal disease, *J. Dent. Res.* **49**:203–222.

Socransky, S. S., 1977 Microbiology of periodontal disease–present status and future considerations, *J. Periodontol.* **48**:497–504.

Socransky, S. S., and Gibbons, R. J., 1965, Required role of *Bacteroides melaninogenicus* in mixed anaerobic infections, *J. Infect. Dis.* 115:247–253.

Socransky, S. S., and Manganiello, S. D., 1971, The oral microbiota of man from birth to senility, *J. Periodontol.* 42:485–494.

Socransky, S. S., Loesche, W. J., Hubersak, C., and MacDonald, J. B., 1964, Dependency of *Treponema microdentium* on other oral organisms for isobutyrate, polyamines and a controlled oxidation–reduction potential, *J. Bacteriol.* 88:200–209.

Socransky, S. S., Manganiello, A. D., Propas, D., Oram, V., and van Houte, J., 1977, Bacteriological studies of developing supragingival dental plaque, *J. Periodontal Res.* 12:90–106.

Socransky, S. S., Sasaki, S., and To, L., 1978, "Piggyback" hypothesis of subgingival colonization of non-motile organisms. II. Migration in or on agar, *J. Dent. Res. Spec. Iss. A* 57:Abstract No. 969.

Sonju, T., and Glantz, P. O., 1975, Chemical composition of salivary integuments formed *in vivo* on solids with some established surface characteristics, *Arch. Oral Biol.* 20:687–691.

Springer, G. F., and Horton, E. H., 1969, Blood group isoantibody stimulation in man by feeding blood group active bacteria, *J. Clin. Invest.* 48:1280–1291.

Staat, R. H., and Schachtele, C. F., 1974, Evaluation of dextranase production by the cariogenic bacterium *Streptococcus mutans, Infect. Immun.* 9:467–469.

Staat, R. H., Gawronski, T. H., and Schachtele, C. F., 1973, Detection and preliminary studies on dextranase-producing microorganisms from human dental plaque, *Infect. Immun.* 8:1009–1016.

Stephan, R. M., 1940, Changes in hydrogen-ion concentration on tooth surfaces and in carious lesions, *J. Am. Dent. Assoc.* 27:718–723.

Stephan, R. M., 1944, Intra-oral hydrogen-ion concentrations associated with dental caries activity, *J. Dent. Res.* 23:257–266.

Stephan, R. M., and Hemmens, E. S., 1947, Studies of changes in pH produced by pure cultures of oral microorganiams, *J. Dent. Res.* 26:15–41.

Stiles, H. M., Loesche, W. J., and O'Brien, T. C. (eds.), 1976, *Microbial Aspects of Dental Caries, Microbiol, Abstr. Spec. Suppl.* 1-3.

Stralfors, A., 1948, Studies on the microbiology of caries. II. The acid fermentation in dental plaque *in situ* compared with lactobacillus counts, *J. Dent. Res.* 27:576–581.

Stralfors, A., 1950, Investigations into the bacterial chemistry of dental plaques, *Odontol. Tidskr.* 58:155–341.

Streckfuss, J. L., Smith, W. N., Brown, L. R., and Campbell, M. M., 1974, Calicification of selected strains of *Streptococcus mutans* and *Streptococcus sanguis, J. Bacteriol.* 120:502–506.

Sumney, D. L., and Jordon, H. V., 1974, Characterization of bacteria isolated from human root surface carious lesions, *J. Dent. Res.* 53:343–351.

Sundquist, G., 1976, Bacteriologic studies of necrotic dental pulps, Ph.D. Thesis, Umea University, Umea, Sweden.

Sundquist, G., and Carlsson, J., 1974, Lactobacilli of infected dental root canals, *Odontol. Revy* 25:233–258.

Svanberg, M., and Loesche, W. J., 1977, The salivary concentration of *Streptococcus mutans* and *Streptococcus sanguis* and their colonization of artificial tooth fissures in man, *Arch. Oral Biol.* 22:441–447.

Svanberg, M. L., and Loesche, W. J., 1978a, Implantation of *Streptococcus mutans* on tooth surfaces in man, *Arch Oral Biol.* 23:551–556.

Svanberg, M. L., and Loesche, W. J., 1978b, Intra-oral spread of *Streptococcus mutans* in man, *Arch. Oral Biol.* 23:557–561.

Swindlehurst, C. A., Shah, H. N., Parr, C. W., and William, R. A. D., 1978, Sodium dodecyl sulphate–polyacrylamide gel electrophoresis of polypeptides from *Bacterioides melaninogenicus*, *J. Appl. Bacteriol.* 43:319–324.

Syed, S. A., and Loesche, W. J., 1978, Bacteriology of human experimental gingivitis: Effect of plaque age, *Infect. Immun.* 21:821–829.

Syed, S. A., Loesche, W. J., Pape, H. L., and Grenier, E., 1975, Predominant cultivable flora isolated from human root surface carious plaque, *Infect. Immun.* 11:727–731.

Takazoe, I., 1961, Study on the intracellular calcification of oral aerobic leptotrichia, *Shika Gakuho* 61:394–401.

Takazoe, I., and Nakamura, T., 1965, The relationship between metachromatic granules and intracellular calcification of *Bacterionema matruchotii*, *Bull. Tokyo Dent. Coll.* 6:29–42.

Takazoe, I., Takeuchi, T., and Nakamura, T., 1963, A chemical investigation of the intracellular calicification of *Bacterionema matruchotii*, *Bull. Tokyo Dent. Coll.* 4:61–75.

Takazoe, I., Matsukubo, T., and Katow, T., 1978, Experimental formation of "corn cob" *in vitro*, *J. Dent. Res.* 57:384.

Tanzer, J. M., Krichevsky, M. I., and Keyes, P. H., 1969, The metabolic fate of glucose catabolized by washed stationary phase caries-conducive streptococcus, *Caries Res.* 3:167–177.

Tanzer, J. M., Chassy, B. M., and Krichevsky, M. I., 1972, Sucrose metabolism by *Streptococcus mutans* SL-1, *Biochim. Biophys. Acta* 261:379–387.

Tanzer, J. M., Hageage, G. J., and Lamon, R. H., 1973, Variable experiences in immunization of rats against *Streptococcus mutans*-associated dental caries, *Arch. Oral Biol.* 18:1425–1439.

Tanzer, J. M., Freedman, M. L., Woodiel, F. N., Eifert, R. L., and Rinehimer, L. A., 1976, Association of *Streptococcus mutans* virulence with synthesis of intracellular polysaccharide, in: *Microbial Aspects of Dental Caries* (H. M. Stiles, W. J. Loesche, and T. C. O'Brien, eds.), *Microbiol. Abstr. Spec. Suppl.* 3:597–616.

Taubman, M. A., and Smith, D. J., 1974, Effects of local immunization with *Streptococcus mutans* on induction of salivary immunoglobulin A antibody and experimental dental caries in rats, *Infect. Immun.* 9:1078–1091.

Taubman, M. A., and Smith, D. J., 1977, Effects of local immunization with glucosyl transferase fraction from *Streptococcus mutans* on dental caries in rats and hamsters, *J. Immunol.* 118:710–720.

Theilade, E., Wright, W. H., Jensen, S. B., and Löe, H., 1966, Experimental gingivitis in man. II. A longitudinal clinical and bacteriological investigation. *J. Periodontal Res.* 1:1–13.

Theilade, E., Larson, R. H., and Karring, T., 1973, Microbiological studies of plaque in artificial fissures implanted in human teeth, *Caries Res.* 7:130–138.

Theilade, E., Fejerskov, O., Prachyabrued, W., and Kilian, M., 1974, Microbiological study on developing plaque in human fissures, *Scand. J. Dent. Res.* 82:420–427.

Thott, E. K., Folke, L. E. A., and Sveen, O. B., 1974, A microbiologic study of human fissure plaque, *Scand. J. Dent. Res.* 82:428–436.

Tinanoff, N., Glick, P. L., and Weber, D. F., 1976, Ultrastructure on organic films on the enamel surface, *Caries Res.* 10:19–32.

To, L., Sasaki, S., and Socransky, S. S., 1978, "Piggyback" hypothesis of subgingival colonization of non-motile organisms. I. Migration through liquids, *J. Dent. Res. Spec. Iss. A* 57:Abstract No. 968.

Tyler, J. E., 1971, Quantitative estimation of volatile fatty acids in carious enamel by gas chromatography of their methyl esters, *J. Dent. Res.* **50**:1189.

van der Hoeven, J. S., 1974, A slime producing microorganism in dental plaque of rats selected by glucose feeding. Chemical composition of extracellular slime elaborated by *Actinomyces viscosus* Nyl, *Caries Res.* **8**:193–210.

van der Hoeven, J. S., 1976, Carbohydrate metabolism by *Streptococcus mutans* in dental plaque in gnotobiotic rats, *Arch. Oral Biol.* **21**:431–434.

van der Hoeven, J. S., and Rogers, A. H., 1978, Discussion of microbial interactions and modification of the microflora, in: *Methods of Caries Prediction* (Workshop Proceedings) (B. G. Bibby and R. J. Shern, eds.), *Microbiol. Abstr. Spec. Suppl.* pp. 119–126, Information Retrieval, Inc., Washington, D.C.

van der Hoeven, J. S., Mikx, F. H. M., Plasschaert, A. J. M., and Konig, K. G., 1972, Methodological aspects of gnotobiotic caries experimentation, *Caries Res.* **6**:203–210.

van der Hoeven, J. S., Mikx, F. H. M., Konig, K. G., and Plasschaert, A. J. M., 1974, Plaque formation and dental caries on gnotobiotic and SPF Osborne–Mendel rats associated with *Actinomyces viscosus*, *Caries Res.* **8**:211–223.

van der Hoeven, J. S., Vogels, G. B., and Bekkers, M. F. J., 1976, A levansucrase from *Antinomyces viscosus*, *Caries Res.* **10**:33–48.

van der Hoeven, J. S., Toorop, A. I., and Mikx, F. H. M., 1978, Symbiotic relationship of *Veillonella alcalescens* and *Streptococcus mutans* in dental plaque in gnotobiotic rats, *Caries Res.* **12**:142–147.

van Houte, J., 1964, Relationship between carbohydrate intake and polysaccharide-storing microorganisms in dental plaque, *Arch. Oral Biol.* **9**:91–93.

van Houte, J., 1976, Oral bacterial colonization: Mechanism and implications, in: *Microbial Aspects of Dental Caries* (H. M. Stiles, W. J., Loesche, and T. C. O'Brien, eds.), *Microbiol. Abstr. Spec. Suppl.* **1**:3–32.

van Houte, J., and Jansen, H. M., 1968a, The iodophilic polysaccharide synthesized by *Streptococcus salivarius*, *Caries Res.* **2**:47–56.

van Houte, J., and Jansen, H. M., 1968b, Levan degradation by streptococci isolated from human dental plaque, *Arch. Oral Biol.* **13**:827–830.

van Houte, J., and Jansen, H. M., 1970, Role of glycogen in the survival of *Streptococcus mitis*, *J. Bacteriol.* **101**:1083–1085.

van Houte, J., and Saxton, C. A., 1971, Cell wall thickening and intracellular polysaccharide in microorganisms from dental plaque, *Caries Res.* **5**:30–43.

van Houte, J., and Backer Dirks, O., de Stoppelaar, J. D., and Jansen, H. M., 1969a, Iodophilic polysaccharide-producing bacteria and dental caries in children consuming fluoridated and non-fluoridated drinking water, *Caries Res.* **3**:178–189.

van Houte, J., Winkler, K. C., and Jansen, H. M., 1969b, Iodophilic polysaccharide synthesis, acid production and growth in oral streptococci, *Arch. Oral Biol.* **14**:45–61.

van Houte, J., de Moor, C. E., and Jansen, H. M., 1970, Synthesis of iodophilic polysaccharide by human oral streptococci, *Arch. Oral Biol.* **15**:263–266.

van Houte, J., Aassenden, R., and Peeble, T. C., 1978, Oral colonization of *Streptococcus mutans* in human subjects with low caries experience given fluoride supplements from birth, *Arch. Oral Biol.* **23**:361–366.

van Palenstein Helderman, W. H., 1975, Total viable count and differential count of *Vibrio* (*Campylobacter*) *sputorum*, *Fusobacterium nucleatum*, *Selenomonas sputigena*, *Bacteroides ochraceus* and *Veillonella* in the inflamed and non-inflamed human gingival crevice, *J. Periodontal Res.* **10**:294–305.

van Palenstein Helderman, W. H., and Rosman, I., 1976, Hydrogen-dependent organisms from the human gingival crevice resembling *Vibrio succinogenes, Antonie van Leeu-wenhoek J. Microbiol. Serol.* **42**:107–118.

Waldman, R. H., and Gangully, R., 1975, Cell-mediated immunity and the local immune system, in: *The Immune System and Infections Diseases* (F. Milgram and E. Neter, eds.), pp. 334–346, Karger, Basel.

Walker, G. J., 1972, Some properties of a dextranglucosidase isolated from oral streptococci and its use in studies on dextran synthesis, *J. Dent. Res.* **51**:409–414.

Walker, G. J., and Pulkownik, A., 1973, Degradation of dextrans by an α-1,6,glucan glu-canohydrolase from *Streptotoccus mitis, Carbohydr. Res.* **29**:1–14.

Wasserman, B. H., Mandel, I. D., and Levy, B. M., 1958, *In vitro* calcification of dental calculus, *J. Periodontol.* **29**:144–147.

Weerkamp, A., Vogels, G. D., and Skotnicki, M., 1977, Antagonistic substances produced by streptococci from human dental plaque and their significance in plaque ecology, *Caries Res.* **115**:245–256.

Weiss, S., King, W. J., Kestenbaum, R. C., and Donohue, J. J., 1965, Influence of various factors on polysaccharide synthesis in *S. mitis, Ann. N.Y. Acad. Sci.* **131**:839–850.

Whiteley, H. R., and Ordal, E. J., 1957, Fermentation of alpha keto acids by *Micrococcus aerogenes* and *Micrococcus lactilyticus, J. Bacteriol.* **74**:331–336.

Wicken, A. J., and Knox, K. W., 1975, Lipoteichoic acids, a new class of bacterial surface antigen, *Science* **187**:1161–1167.

Williams, B. L., Pantalone, R. H., and Sherris, J. C., 1976, Subgingival microflora and peri-odontitis. *J. Periodontal Res.* **11**:1–18.

Williams, J. L., 1897, A contribution to the study of pathology of enamel, *Dent. Cosmos* **39**:269–353.

Williams, R. A. D., 1967, The growth of Lancefield group D streptococci in the presence of sodium fluoride, *Arch. Oral Biol.* **12**:109–117.

Williams, R. A. D., and Shah, H. N., 1979, Enzyme patterns in bacterial classification and identification, in: *Impact of Modern Methods on The Taxonomy of Bacteria*, Soc. *Appl. Bacteriol. Symp. Ser.* (in press).

Williams, R. A. D., Bowden, G. H., Hardie, J. M., and Shah, H. N., 1975, Biochemical prop-erties of *Bacteroides melaninogenicus* subspecies *Int. J. Syst. Bacteriol.* **25**:298–300.

Wilton, J. M. A., 1977, The function of complement in crevicular fluid, in: *The Borderland between Caries and Periodontal Disease* (T. Lehner, ed.), pp. 223–247, Academic Press, London.

Wilton, J. M. A., Ivanyi, L., and Lehner, T., 1971, Cell-mediated immunity and humoral antibodies in acute ulcerative gingivitis, *J. Periodontal Res.* **6**:9–16.

Wilton, M., 1969, A comparative study of circulating and cell-mediated immunity induced by the gingival and systemic administration of oral bacteria, *J. Dent. Res.* (*Abstr. British Division IADR*) **48**:1098.

Wittenberger, C. L., and Angelo, N., 1970, Purification and properties of a fructose-1,6-diphosphate-activated lactate dehydrogenase from *Streptococcus faecalis, J. Bacteriol.* **101**:717–724.

Wittenberger, C. L., Palumbo, M. P., Bridges, R. B., and Brown, A. T., 1971, Mechanisms for regulating the activity of constitutive glucose degradative pathways in *Streptococcus faecalis, J. Dent. Res.* **50**:1094–1102.

Wolin, M. J., 1964, Fructose-1,6 diphosphate requirement of streptococcal lactic degydro-genases, *Science* **146**:775–777.

Wood, J. M., 1964, Polysaccharide synthesis and utilization by dental plaque, *J. Dent. Res.* **43**:955.

Wood, J. M., 1967, The amount, distribution and metabolism of soluble polysaccharides in human dental plaque, *Arch. Oral Biol.* **12**:849–858.

Wood, J. M., 1969, The state of hexose sugar in human dental plaque and its metabolism by the plaque bacteria, *Arch. Oral Biol.* **14**:161–168.

Woolley, L. H., and Rickles, N. H., 1971, Inhibition of acidogenesis in human dental plaque *in situ* following the use of topical sodium fluoride, *Arch. Oral Biol.* **16**:1187–1194.

Yamada, T., and Carlsson, J. C., 1975, Regulation of lactate dehydrogenase and change of fermentation products in streptococci, *J. Bacteriol.* **124**:55–61.

Yamada, T., and Carlsson, J. C., 1976, The role of pyruvate formate-lyase in glucose metabolism of *Streptococcus mutans*, in: *Microbial Aspects of Dental Caries* (H. M. Stiles, W. J. Loesche, and T. C. O'Brien, eds.), *Microbiol. Abstr. Spec. Suppl.* **3**:809–819.

Yamada, T., Hojo, S., Kobayashi, K., Asano, Y., and Araya, S., 1970, Studies on the carbohydrate metabolism of cariogenic *Streptococcus mutans* strain PK-1, *Arch. Oral Biol.* **15**:1205–1217.

Zander, H. A., Hazen, S. P., and Scott, D. B., 1960, Mineralization of dental calculus, *Proc. Soc. Exp. Biol. Med.* **103**:257–260.

Index